Chasing the Wind

Other Books by the Author

Different Battles: The Search for a World War II Hero
In Their Footsteps: Warriors, Capitalists, and Politicians of West Virginia
The Rise and Fall of Dodgertown: 60 Years of Baseball in Vero Beach

Chasing the Wind

Inside the Alternative Energy Battle

Rody Johnson

The University of Tennessee Press • Knoxville

The paper in this book meets the requirements of American National
Standards Institute / National Information Standards Organization specification
Z39.48-1992 (Permanence of Paper). It contains 30 percent post-consumer waste
and is certified by the Forest Stewardship Council.

Library of Congress Cataloging-in-Publication Data

Johnson, Rody.
Chasing the wind: inside the alternative energy battle / Rody Johnson. — First edition.
pages cm
Includes bibliographical references and index.
ISBN 978-1-62190-029-0 (hardcover)
1. Wind power.
I. Title.
TJ820.J66 2014
333.9'2—dc23
2013039693

To Ruth Baker Thomas Johnson (Tommye)

Contents

Illustrations

FIGURES

TABLES

Prologue

Arriving like a summer storm, wind energy became an issue both where I live in Florida and where I spend my summers in West Virginia. In the winter of 2006, I began to notice a series of letters to the editor in the weekly newspaper serving Greenbrier County, West Virginia. The letters objected to plans to install a huge wind farm on nearby mountain ridges. My first exposure to wind energy had come years earlier, upon seeing the conglomeration of windmills in a desert valley while driving east from Los Angeles. It was an interesting sight, rather unique, but in my mind just one of those California things. I was amazed to learn that a wind farm already existed in West Virginia and even more so that wind turbines might appear on Florida beaches.

In Lewisburg, West Virginia, I found a battle that was tearing a community apart. The situation dominated the local news. I talked with people for and against the wind farm, went to meetings and hearings, and visited the heads of the opposition groups. In time, I made contact with the wind farm developer to get his perspective. Initially I developed biases against wind energy based on the local situation, since it seemed to do little good for the community and was so divisive. The local politics added to the conflict. Meanwhile, with Al Gore leading the way, the climate change issue became popular, though I wasn't sure I believed in all that was being said. I wanted to get an understanding of that situation as well.

Then wind energy surfaced as an issue near my home in Florida. NextEra Energy (formerly FPL Group), a corporation that owned a vast percentage of the wind energy capacity in the United States, planned to place turbines on the beach near one of its nuclear power plants. Thus another battle began.

As I expanded my research, I learned about the history of wind power, sought out the experiences that were being had in other parts of the United States, and studied what was going on in Europe—the early world leader in the use of renewable resources. My thinking about wind began to broaden. Wind had its place. But was it going to lead the way in reducing the world's greenhouse gas emissions, as one gathers from the developers who will profit from it, from the media who report on it, and from the politicians who use it to build their careers? Does it justify the government subsidies it receives?

I needed to expand my perspective and look at other energy resources besides wind. Coal was king in West Virginia, but in Florida the full breath of resources were being explored—coal has been rejected, but natural gas, nuclear, solar, and wind energy were active considerations in the growing demand for power. And what about conservation?

In northern West Virginia, I visited an established wind farm and talked with people who lived nearby. In Greenbrier County I watched a wind farm transform miles of ridgetop. In Florida I took a tour of a planned beachfront wind farm and visited the nearby nuclear power plant. And during all of this, I continued to talk with people on both sides of the issue and read everything I could find. I also attended a forum on wind energy.

My sources included the few books that have been written specifically on the subject, many current newspaper and magazine articles, various technical and economic reports, and the Internet. The latter provided a wide breath of information, some of it opinionated. The subject is controversial and, therefore, searching for the truth was challenging.

This book is the story of wind energy as I have discovered it. I have tried to leave my biases behind and to present the issues so that readers can draw their own conclusions. The issue: Should we believe in wind energy as a significant solution to preventing climate change?

I am grateful for the communications I have had with Dave Groberg, who led the development of the Greenbrier County wind farm, and the conversations with John Stroud and Debbie Sizemore who led the effort to oppose it. Friends sent me information, particularly Dennis Moloney, who also drove me in his pickup truck on visits to the Greenbrier wind sites. My son, Mark Johnson, a newspaper writer, reviewed my early work and attempted to put me on the right path. Reed Johnson, former professor of nuclear engineering at the University of Virginia, advised me on the intricacies of nuclear power. Phil Sparks, the retired senior manager of state and local affairs at Dominion Resources, reviewed the manuscript of this book, as did Carter Taylor, an energy conservation consultant. John Byram, formerly of the University Press of Florida and now director of the University of New Mexico Press, first recognized the book's potential. The advice of peer reviewers at the University Press of Florida and the University of Tennessee Press improved its structure. Sian Hunter provided council on seeking publication. Kerry Webb guided me through the University of Tennessee Press's acquisition process. The copyediting by UT Press's Gene Adair and freelancer Karin Kaufman made me appear a better writer than I am. And finally, my wife, Tommye spent hours reading chapters and correcting errors that I could never see.

Chapter 1

From the Beach to the Mountains

OCEANFRONT CONFRONTATION

On a sunny winter day on the east coast of Florida, three buses headed down State Road A1A south of Fort Pierce. Aboard were a group of local citizens and the press, who had been invited to inspect a site where Florida Power and Light proposed to place wind turbines. The location was a narrow strip of island bordered by the Atlantic Ocean on one side and the Indian River Lagoon on the other. A bit farther south on this same island sat, most conspicuously, two round containment buildings of the power company's St. Lucie Nuclear Power Plant.

On the ride to the site, a tall, stylish woman stood at the front of the bus. She had the looks of a high-end real estate saleswoman, but she introduced herself as Henrietta McBee, the power company's director of project development. When later asked, she would reveal that she had both engineering and MBA degrees. As the bus rumbled along, McBee talked about the project and answered questions from the passengers. There would be nine turbines, she said, four on public property and five at the nuclear power plant, spread along six miles of beachfront. They would provide enough electricity to power on average twenty-eight hundred homes.[1] The turbines would start generating electricity when the wind reached a level of seven to ten miles per hour and could operate in wind up to fifty-five miles per hour. When she began to talk about the height of the turbines—four hundred feet—someone in the back of the bus mentioned that they would be about twice as high as the twenty-story condominiums they had just passed. Henrietta quickly moved on, covering such technical details as the blade pitch, the megawatt capacity of each turbine, the turbines' life span, and the fact that there would be no need for a transmission line as lines for the nuclear plant were already available.

The buses slowed and turned into the dirt parking lot of a scrubby beachside park. The passengers stepped out and, to their surprise, faced a circle of protesters chanting slogans and holding signs. McBee, taken back, rounded up her charges and walked them to a large square area marked off by red tape. The protesters joined the crowd. She explained to the visitors that this would be where the base of the turbine would be located, sitting just inland from the dune line and not more than one hundred feet from the beach. The area was bare except for ankle-high weeds that the crowd found cumbersome to wade through. The turbine towers, McBee noted, would be mounted on concrete pedestals that would resist a possible wave surge generated by a hurricane. Across the dunes, a slight breeze rippled the ocean. Was this enough wind to start blades turning? Trying to imagine a four-hundred-foot wind turbine towering over this site was difficult.

The protesters had now mixed with the bus passengers. They included retirees, homeowners, and environmentalists, holding their ragtag signs high for the local TV news cameras. One man carrying a sign said he had copied it from an antiturbine logo found on the Internet and then pasted it on cardboard. He lived on a street of riverfront homes that would have a direct view of the turbines across the mile-wide lagoon. The protesters did not seem to represent any one organized group and were generally polite. McBee talked about turbine noise, saying, "If I was standing under a turbine right here, I could be carrying on a conversation just like now." In response, a man standing nearby made a sound—*swoosh, swoosh, swoosh, swoosh*—briefly providing a background noise that interrupted her. She took it calmly, listened to concerns, and answered questions and comments.

"There is not much wind in Florida. Why put them here?"

"What would be the impact of a hurricane?"

"We are already staring at the nuclear plant."

"What would be the effect on sea turtles, or on spawning sea trout?"

"What about a storm surge in a category 4 or 5 storm?"

"This is an extremely environmentally sensitive area. You must do an environmental impact study first."

McBee responded at times with "that's a great question," answering it fully. When the questions were tough, she fell back on environmental benefits and climate change themes. With regard to the hurricane impact, she said that the turbines stood in Tornado Alley (in the Midwest). She agreed that Florida was not a very windy state but said that wind energy would supplement the power from conventional plants, reducing that power "as we bring up" the wind turbine power. The coast was the only area with suffi-

From the Beach to the Mountains

Florida Power and Light's Henrietta McBee (far left) presents the facts about a planned wind farm along the beach front near Fort Pierce, Florida. (Photograph by the author.)

cient wind for the turbines, and there was not a lot of space available due to condos and development. It would be "much more difficult" to place turbines offshore. And yes, environmental impact studies would be done.

As the buses reloaded to go down the beach to another turbine site and the protesters followed in their cars, McBee talked a bit more frankly to the people sitting near her. She said that Florida Power and Light had looked at the area above Cape Canaveral and the Kennedy Space Center farther up the coast, which had better winds, as a possible site. The U.S. Air Force and NASA had no objections, she said, but the Fish and Wildlife Service did. That agency managed the Canaveral National Seashore and the Merritt Island Wildlife Refuge. "I started out with eighteen wind turbines for the Florida project, and now I'm down to nine," she stated. She called the project "weird" in that it was unlike any she had worked on in the past. Florida Power and Light would have to get federal approval to put turbines on the nuclear power plant property. In referring to one of her past projects, she mentioned the Mountaineer Energy Center facility in Tucker County, West Virginia, owned by NextEra Energy Resources (formally FPL Energy), a sister company of Florida Power and Light, and enthusiastically described her husband's reaction when he

3

first saw the turbines there, "coming over a rise in the highway at a height where you were looking head on at the blades."

AN OPERATING ATTRACTION

It fills the front windshield. Anyone driving north on winding U.S. 219 who crosses the neck of West Virginia's eastern panhandle in Tucker County has the experience. As a car comes up over a knoll, its driver sees three enormous blades. They look like the propeller of a huge aircraft that may have crashed. The instinct is to dodge, but the highway quickly curves downward to the right. A panoramic view of ten gigantic wind turbines lined up along a bare mountain ridge then appears out the side window. Alongside the highway, cars park in a pull-off and people get out to see the sight. No more than fifty yards away stands the base of the first turbine, a massive structure. Here lies an industrial tourist attraction.[2]

On a sweltering summer afternoon with no breeze at ground level, the blades far above grabbed a bit of wind and turned slowly, majestically. Despite the noise of cars whizzing by on the highway, the low, slow *thump, thump, thump* from the rotating blades sounded like a heavily loaded washing machine.

A pickup truck pulled up and parked. A technician got out and in response to a question announced that one of the turbines had a gearbox problem that he had to check. He unlocked a gate and drove down a road alongside the towers. The blades on one of the far turbines stood motionless.

A tour bus rolled into the parking lot. An elderly crowd eased their way off the bus with cameras in hand and stood gawking at the turbines like a group of tourists viewing the Golden Gate Bridge. Other cars pulled off the highway; their passengers got out and looked. When they got home, each would have a story to tell of the sight they had seen.

Local residents in the nearby Tucker County communities of Thomas and Davis generally favored the wind energy facility. The area, once dependent on coal and timber, had a few motels and restaurants serving people visiting the nearby Blackwater Falls State Park and the Canaan Valley ski resorts. Judy Lambert, who lived within a mile of the turbines and who didn't seem particularly bothered by the noise, said, "If the blades are facing us, we hear them the most. . . . If it's really windy, you can hear them moving very fast." Though the windmills can be seen from the tops of the mountains above Canaan Valley some twelve miles distant, realtor Laura Reed said that real estate sales had been "constantly increasing at a steady rate." She went on

4

Drivers in Tucker County, West Virginia, approaching the Mountaineer wind farm, see wind turbines face to face. (Photograph by Gary Cooper.)

to say that "the windmills had become a tourist attraction. They're an odd-ity." On the other hand, one fellow said, "They are ugly eyesores. They're just not natural-looking the way they stick out of the tallest mountain and you can see them from so far away." But John Bright, owner of a coffee shop and market in Thomas, took a broader view: "I grew up in this state watching the land get raped and polluted with orange water and acid rain. Any form of energy is better than the coal mines." David Downs, who owned an antique shop and a real estate agency, said the turbines that were already there were all right since they were in an area that didn't bother too many people. How-ever, the only thing West Virginia had going for it was its scenery. And here he got a bit emotional. To make any difference, that is to produce enough power to do any good, he said they were going to have to put turbines on every mountain ridge in the state. That would be "monstrous," he said.

At the Tucker County Convention and Visitors Bureau in nearby Davis, executive director Bill Smith and a couple of his staff were in the midst of sorting pamphlets to put into tourist packages. When asked about the wind-mills, he indicated that the turbines had a "high visual impact." But from the area's two major tourist attractions they could be seen only from the top of the ski lifts in Canaan Valley (twelve miles away) and only from the road going into Blackwater Falls. (From that location on a slightly hazy afternoon, twenty-five turbines were visible, lined up on Backbone Mountain some four to five miles in the distance.) He noted that in plain sight of the turbines, property values could decline.

Bill Smith provided a visitor's guide and an FPL (now NextEra) Energy fact sheet. The Mountaineer Energy Center consists of forty-four turbines, the guide noted, stretching along six miles of a ridge known as Backbone Mountain. The turbines measured 350 feet in height and if placed next to the Statue of Liberty in New York Harbor would be slightly higher.

In the back of the visitor's guide among the advertisements, a picture of two wind turbines with wildflowers in the foreground was captioned, "Got Power? Tucker County's wind harvest is WV's first and only alternative energy source." The FPL Energy fact sheet declared, "West Virginia's vast coal reserves have helped power America's industrial might for more than two centuries. . . . West Virginia is now tapping another of its abundant natural resources to help fuel the nation and power its own economy—the wind." The fact sheet explained how the operation worked: "The wind moves the blades just as a child blowing on a pinwheel makes the pinwheel turn. The windmill blades turn a shaft inside a generator to make electricity." Exelon Generation Company purchases from NextEra all the power being generated on a long-term contract and distributes it to Pennsylvania, Maryland, and the Washington, D.C., area. The sheet touted the benefits of wind energy, noting that it "generates clean power to be used in the region; requires little land and surrounding land can be used for other purposes; provides tax payments to local governments; provides lease payments to landowners where turbines are installed; and places little or no burden on local infrastructure, such as public schools, services."

At the Saw Mill restaurant in Davis, waitress Jenny Johnson had her say. She lived within a mile of the turbines. They didn't bother her; the noise didn't bother her. But a couple of her neighbors "detested" them. To her, however, it was like living in the city or near train tracks or an airport—you got used to the noise. She was so attuned to the sound that she could tell when the turbines swung with a change in wind direction. Once she heard a banging and called one of the wind farm technicians she knew. The noise turned out to be a loose door at the bottom of the tower swinging in the wind.

No one mentioned the bat kill issue. But to some environmentalists, it is a serious concern. An early study at the site concluded that the forty-four tur-

From the Beach to the Mountains

bines had killed two thousand bats during a six-week migration period. The West Virginia Highlands Conservatory was particularly concerned about the high probability of bat kills from wind farms located on the heavily forested Appalachian Mountains.

Another issue came up in nearby Parsons at the one-room storefront office of the *Advocate*, Tucker County's weekly newspaper. A young couple, Kelly and Chris Stadelman, owned the paper. Kelly Stadelman said the only controversy with the Mountaineer wind farm that she could recall was the property tax situation. She felt that the county commission had been so excited when the wind facility came that they had not negotiated a satisfactory deal for the county. Four West Virginia state legislators, whose districts covered the area where wind farms were being considered, had pushed through a bill in 2001 that allowed property taxes to be based on a salvage value of 5 percent for wind projects rather than the normal 60 percent applied to other enterprises. NextEra officials, among them Henrietta Bee, had come to Tucker County to negotiate the property tax with the county commissioners. The commissioners wanted a $300,000 contribution per year for twenty years; NextEra offered $35,000 and settled for something closer to $100,000. County commission president Sam Eichelberger said, "I just think for what they're getting on the dollar out there they could treat our county better." McBee responded, saying, "Without tax breaks, including one at the federal level, wind power projects would not be feasible."[3] She also said at the Florida site that NextEra wouldn't build in West Virginia again. They couldn't "cope with the politics and the coal influence."

Meanwhile, on similar ridges over one hundred miles south in Greenbrier County, West Virginia, the sound of heavy equipment echoed across the mountaintops as a bigger wind farm was under construction.

MOUNTAINTOP RENOVATION

Dennis Moloney and his wife, an attorney, who own a seventy-five-acre farm in the valley on Chestnut Flats about eight miles outside the Greenbrier County seat of Lewisburg, moved from Washington, D.C., to this area during the Back to the Land movement of the 1970s, seeking a quieter life. From a hill on his property, he could see the ridge where Invenergy, a Chicago-based wind-farm developer, was constructing wind turbine sites. They had leased the land from MeadWestvaco, a paper company that had cut timber on the property for a hundred years. The operation was called Beech Ridge Energy.

7

On an August day in 2009, Dennis decided to see what was happening on the ridgetops. In his Dodge Ram pickup truck, he drove several miles across the valley to the base of the ridge. He followed a narrow gravel road that wound upward through thick strands of towering trees to the top of Cold Knob, at four thousand feet the highest point in Greenbrier County. Coming off the peak, Dennis flagged down a logger driving a tractor-trailer and asked if he had seen the construction. The logger pointed ahead. At an intersection of several gravel roads, a sign indicated a work safety zone and an arrow pointed ahead to a string of turbine sites. Dennis could now hear the rumbling sounds of equipment. The first site sat off to the left on the side of Cold Knob, with a wide, sweeping gravel road heading up the ridge. Some distance away, the base looked finished and already seeded with grass. Next a sign pointed to a location where both the sites and the road were under construction. A pickup truck whizzed by. A front-end loader sat by the road. Trucks hauling gravel moved past, spewing dust. Dennis drove on along the string of sites. While most of the concrete bases were completed, others had foundations that were not yet covered with dirt.[4]

According to the signs, there were still many sites ahead, but Dennis turned around and retraced his route, looking for other strings on other ridges. He came to a road that was more developed, at least three lanes wide, well graveled, possessing culverts for drainage, and seeded on the edges. It would take roads like this to haul tower sections and the long blades up the mountain to the sites. This road ran along the top of a ridge and reminded him of the Skyline Drive in Virginia, which followed the crest of the Blue Ridge Mountains. He was above the clouds that at noon still lay in the valleys below. A site was close to the road. He stopped and took a look at the foundation. Made of poured cement, the lower (to be buried) portion was octagonal, rising to a circular mount, protruding above ground, and had closely placed studs around the circumference where the turbine tower would be mounted. The foundation and its protruding base give the appearance of a spaceship that had landed. Dennis Moloney continued past a string of fifteen sites lined up along the road. At one point he saw a hand-lettered sign marking a helicopter landing area.

On another, busier road marked for a different string of sites, cable reels and power poles were staged with what appeared to be a foundation for a small building, perhaps the site's substation. A crane had installed a power pole for a transmission line and was raising another. Dennis stopped and talked to a trucker, who said that the road led to a specially built cement plant that provided the concrete for the tower foundations.

8

Dennis returned home after seeing only a portion of this whole complex. This was not mountaintop removal, as in coal mining, but from the air the scars from the construction must have looked like a giant octopus with its head on Cold Knob and multiple tentacles reaching out along the ridges in all directions. Back in the valley near his farm, Dennis looked at the distant ridge where he had just been and where in several months a string of four-hundred-foot wind turbines would be perched. It had been four years since this project was first proposed—four years of controversy.

BLADES, TOWERS, AND NACELLES

Beech Ridge, when completed, would be a typical wind farm, except that it would lie in mountainous terrain rather than on the plains of the Midwest, where most wind facilities were located. Work on the Beech Ridge site had begun a few months before Dennis's visit. The scrub growth left over from strip mining and timber cuts had been cleared for the gravel roads and the turbine sites. The turbines were to be grouped into clusters or strings, depending on the terrain, and accessed by the road network. The steel and concrete foundations for the towers had been completed for the most part. The turbine foundation that Dennis inspected was a fifty-foot-wide octagonal footing with a depth of four to six feet with a twenty-foot-wide round pedestal protruding from the center of the footing. The turbine's pedestal is locked into the ground by thirty-foot-deep anchor bolts. The footing for a four-hundred-foot wind turbine is exposed to tremendous loads, especially in high wind, when the rotor is locked down. Because of the possibility of extreme stresses, the base is described as a "mechanical device" rather than a foundation.[5]

With the footings in place, the General Electric (GE) 1.5-megawatt turbine components arrived at the sites on semitruck trailers. The trucks turned off of West Virginia's I-64 onto a two-lane state road, passed through the small town of Rupert on their way up a valley, then climbed gravel roads to the various ridgetops, a trip from the interstate of roughly twenty miles. Each turbine consisted of the three blades, the modular tower sections, and the nacelle. The 120-foot blades were carried on separate, specially designed trailers, and the cylindrical tower sections were hauled on flatbeds, as were the nacelles. By themselves, the nacelles look like house trailers sitting on the flatbeds. The blades and tower sections came from Iowa and Texas, where they were fabricated by GE subcontractors, and GE shipped the nacelle, the heart of the turbine, from its own assembly plants in Greenville, South Carolina, and Pensacola, Florida.

From the Beach to the Mountains

Once on site, crews assemble a turbine in a matter of a couple days using a huge crane. The tower section is mounted to the foundation pedestal, the sixty-ton nacelle is lifted to the top of the tower, and the blades are attached to the nacelle hub at a height of 260 feet.[6] The overall height with the blades in place reaches almost 400 feet.

Underground power cables connect each turbine to the newly built, on-site substation. From there Invenergy constructed a thirteen-mile power line across the ridges to connect the system to a substation and transmission line that was already part of the grid that fed East Coast cities such as Washington, D.C., and Baltimore.

SPINNING THE BLADES

Wind flow across Greenbrier County ridges spins a turbine's blades, which rotates the hub at low speed. Within the nacelle, a gearbox transfers the low-speed rotation to a high-speed shaft that drives the generator. Alternating

A wind turbine foundation under construction at the Beech Ridge site in West Virginia. (Courtesy of Mountain Communities for Responsible Energy.)

Table 1.1. Wind Speed's Impact on Turbine Power and Capacity Factor		
Wind Speed (mph)	**Power (MW)**	**Capacity Factor (%)**
7	0	0
16	0.5	33
22	1.0	66
31–56	1.5	100

Source: GE Energy, 1.5MW Wind Turbine spec sheet.

current (AC) electricity is generated and flows through a power line to the power grid for distribution to many users. (Farm windmills, so popular in the early 1900s, generated direct current [DC], which was stored in batteries in the home prior to use.) In conventional power plants, fossil fuels or a nuclear reaction produces heat and converts heated water to high-pressure steam, which drives the generator.

Computer controls in the nacelle react to changes in wind speed and direction. The controls adjust the blade pitch to the wind speed and face the rotor into the oncoming wind. The blades grab the wind and begin to turn. When the wind speed reaches seven miles per hour, electricity begins to flow. As the wind increases, the power increases—not directly but exponentially. The turbine's maximum output is 1,500 kilowatts or 1.5 megawatts. When the wind speed reaches sixteen miles per hour, the turbine produces 0.5 megawatts (see Table 1.1). A roughly 40 percent increase in the wind speed, from sixteen to twenty-two miles per hour, increases the power output 100 percent, from .5 to 1.0 megawatts. The turbine reaches its maximum power of 1.5 megawatts at a wind speed of thirty-one miles per hour and holds that output up to fifty-six miles per hour, at which time it shuts down to protect the turbine. The power levels off at thirty-one miles per hour, because wind turbines generate energy by "slowing down the wind." A turbine's rotor at this point can only catch 59 percent of the wind that flows through it, thus spilling the rest. This phenomenon is called the Betz Limit.[7]

At thirty-one miles per hour and up, the turbine is operating at its maximum rated capacity of 1.5 megawatts and its capacity factor is 100 percent. At sixteen miles per hour, the turbine is generating 0.5 megawatts versus its

From the Beach to the Mountains

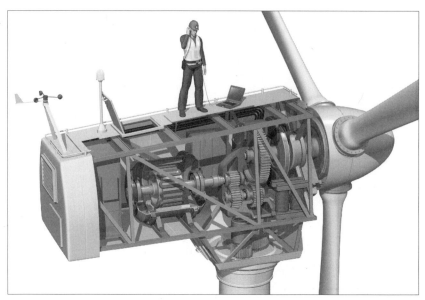

A cutaway view of a wind turbine. Wind spins the blades, turning the rotor and then the shaft, which, through the gear train, spins the generator at high speed, thus producing electricity. (Illustration by Stephen Sweet, Dreamstime.com.)

rated capacity of 1.5 megawatts. Therefore, its *capacity factor* (or efficiency) is 33 percent. Since wind is variable at any instant, a more meaningful term is *average capacity factor,* which is based on a period of time, generally a year. If the wind at the Beech Ridge site averages sixteen miles per hour over a year's time, then the wind farm's annual average capacity factor is 33 percent, though it can vary at any point in time from 0 to 100 percent, depending on the wind speed. Because different regions have different wind potential, average wind farm capacity factors vary. According to the American Wind Energy Association, these factors range from 22 percent in New England to 32 percent in Texas.[8]

Wind developers almost never discuss capacity factors, because they are significantly lower than those of coal, natural gas, and nuclear-powered plants. These developers prefer to talk about the number of homes on average that can be supplied electricity by a wind farm. A 1.5-megawatt wind turbine operating for a year with an average capacity factor of 34 percent can supply on average 370 typical homes. When the wind is under seven

From the Beach to the Mountains

Table 1.2. Wind Farm versus Nuclear Plant: Average Capacity Factor and Homes Served		
Items	**Wind Farm**	**Nuclear**
Plant installed capacity (MW)	100	2,200
Average capacity factor (%)	34	90
Average power output (MW)	34	1,980
Average homes served[a]	25,000	1,400,000

[a]Average homes served is calculated by multiplying the average power output (megawatts x 1,000) times the hours in a month divided by the average kilowatt-hours used per home (1,000 kilowatt hours per month).

Source: Energy Information Administration, "Annual Energy Outlook 2011."

miles per hour, no homes will be supplied, while at thirty-one miles per hour, over 1,000 homes (three times as many) will be supplied. Rather than saying that their wind farm operates at 34 percent efficiency, it sounds more impressive to say that a medium-sized wind farm with an installed capacity of 100 megawatts serves on average twenty-five thousand homes (see Table 1.2). However, a typical nuclear power plant with 2,200 megawatts of installed capacity operating at 90 percent efficiency (average capacity) supplies electricity consistently to 1.4 million homes. In essence, a typical nuclear plant has roughly twenty times more installed capacity than the wind farm but can supply more than fifty times more homes because of its higher capacity.

Since the 1980s and the days of the California wind energy boom, turbine efficiency and output have increased dramatically, thus reducing the cost of wind power. A turbine's rotor (blade) diameter is now almost five times larger. The average rated capacity per turbine has gone from 50 kilowatts to mostly 1,500 kilowatts (the GE 1.5-megawatt unit) and as high as 6,000 kilowatts for offshore units. On land, narrow streets, sharp curves, overpasses, overhanging trees, and traffic conditions can place a limit on the use of large turbines. And while windmills that generate utility-type power are a recent innovation, windmills themselves date back to the Middle Ages and beyond.

From the Beach to the Mountains

Chapter 2

A Thousand Years

FROM PERSIA TO THE AMERICAN WEST

In this the twenty-first century, windmills maybe causing controversy in the United States, but they have been on earth for over a thousand years. In Persia, the Iran and Iraq of today, they existed in the tenth century as a carousel-type structure that ground grain and pumped water for gardens. Perhaps the Crusaders brought back to Europe from the Mideast the windmill concept because they began to appear in England and France as tide mills and water wheels as early as 1100. By 1300, windmills had spread across Europe. The Dutch in particular took to these devices, having at one time ten thousand in their small country. They used them together (the first wind farms) to pump water and to reclaim land from the North Sea. Robert Righter, a wind energy historian, wrote that windmills were the "most complex power device of medieval times," preceding the coming Industrial Revolution.[1] Throughout Europe they pumped water, ground corn and wheat, and were the basic power for sawmills. It was estimated that windmills supplied 25 percent of Europe's power beginning in 1300, with the rest coming from the waterwheel and the labor of men and animals. By the 1800s, the arrival of the steam engine and coal had changed the power equation.

In colonial America, with its many creeks and streams, waterwheels, not windmills, became the principal source of power. However, windmills did exist. Colonists built the first one at Jamestown in 1621. The Dutch, following the trend in their native country, established windmills at New Amsterdam (New York City). With its windy conditions, Cape Cod became an ideal location. Shingled and whitewashed, the windmills there resembled women's smocks and thus were called smock windmills. Most people found them "attractive and colorful," though even then there were some objections. Writer and early environmentalist Henry David Thoreau thought they looked like huge turtles.[2]

But it was in the West, with its dry, windy conditions, that windmills became popular. Water was in short supply and wind supplied the solution, providing the mechanical power to pump water from wells. As the railroads expanded westward after the Civil War, so did windmills. The Union Pacific Railroad bought the first commercially built unit in America. It had a rotor thirty-nine feet in diameter and supplied water for the locomotives' steam engines. The growth of the cattle industry followed, making the windmill and its nearby pond a familiar sight across the plains. Six million windmills, so it was estimated, operated in this area from the late 1800s to the 1930s. Unlike today, they were considered compatible with nature and part of the local landscape.

INNOVATION BUT NO MARKET

By the 1880s American city dwellers were beginning to enjoy the benefits of electricity—lights in their homes, telephones, and electric streetcars for transportation. Coal-fueled central power plants generated the electricity. At the time, wind power was not a consideration. It took Charles Brush, a wealthy Ohio inventor and industrialist, to develop a windmill that could generate electricity. In 1888 he installed in his five-acre backyard a mammoth, multiblade, farm-type unit fifty-six feet in diameter. It fed a group of storage batteries, providing power for one hundred light bulbs, two arc lights, and three electric motors.[3] By 1900 he had abandoned his machine and taken advantage of the electricity supplied by Cleveland's central power plant. He had developed a machine that in America at the time had no market.

The multiblade windmills in the West that pumped water turned too slowly to efficiently generate electricity. Research on propellers during World War I resulted in airplane-type blades being used on windmills. They turned six to ten times faster than the existing units and became popular in the 1920s. Marcellus Jacobs, the founder of the Jacobs Wind Electric Company, took advantage of this technology by reducing the number of blades from twelve to three, using propeller blades, changing their pitch, and getting three times more energy production. His company sold thirty thousand propeller-type units between 1927 and 1957 throughout the Midwest to farms that had no access to power plants, even though the units cost as much as fifteen hundred dollars apiece. At the lower end of the market, the Wind-

A Thousand Years

charger Corporation sold thousands of smaller units that could charge a battery to run a radio. The Sears Roebuck catalog advertised wind energy units for the home and farm. However, the cost of the units and the Depression of the 1930s limited their full market potential. The 1936 passage of the Rural Electrification Act brought electricity to rural areas, and over the next twenty years, the need for farm windmills decreased. As author Peter Asmus commented, "The public power movement all but killed America's wind farming industry."[4]

The Jacobs Wind Electric Company, in hopes of continuing in business, proposed in the 1950s what would have been the first wind farm concept in the United States—one thousand windmills, one mile apart, strung along the power line system that ran from Montana to Minnesota. It would have fed directly into the regional grid of that time.

The ability to feed wind energy into a grid existed because of Palmer Putnam, an inventor and engineer who developed an interest in wind energy when he built a house on Cape Cod, where he found both the winds and the electric rates high.[5] But he was not interested in providing electricity for a home or a farm, but for a town. In 1941, assisted by a 350-man engineering team, he developed and installed a utility-scale windmill, or what would now be called a wind turbine, on a two-thousand-foot-high ridge in the Green Mountains of Vermont. It was a spot where the wind blew at an annual average speed of thirty-eight miles per hour, the highest average speed on the East Coast. This two-bladed unit standing two hundred feet high turned at such a high speed that it did not generate direct current as other windmills but, with a synchronous generator, provided alternating current that could feed into the lines of existing central power systems. A local power company accepted Putnam's proposal to use his turbine to work with a hydro plant, supplying electricity while conserving water power. At wind speeds of thirty miles per hour, it produced 1.0 megawatt of power, and in higher winds, 1.5 megawatts. With the exception of a unit in Russia, no other machine in the world of this size and capability existed. After operating for sixteen months, a bearing failed on Putnam's machine, and because of wartime shortages, it could not be replaced until 1945. Within weeks of going back on line, a blade broke away from the rotor during a 120-mile-per-hour windstorm and the machine never operated again. Like Charles Brush, Palmer Putnam produced a machine that was not only ahead of its time but also would be the prototype for the units of today. However, the machine's blade failure and the potential for hurling ice that build up on its blades were two concerns used by wind energy opponents to question the feasibility of wind turbines.

17

NUCLEAR POWER AND WIND ENERGY

By the 1960s, with the growth of energy doubling every ten years, nuclear power, not wind energy, became the power source of the future. The U.S. Atomic Energy Commission, formed after World War II, invested $27 billion in nuclear energy between 1955 and 1964. Utility companies ordered over two hundred nuclear units. Then, in 1979, the nuclear plant at Three Mile Island failed, causing radiation leakage. The scare, with both an emotional and political impact, resulted in the cancellation of orders for plants, leaving approximately a hundred units operating in the United States to this day. But even this number provided 20 percent of the nation's electricity. In 1986, the Chernobyl nuclear plant in Russia spewed radiation over a large area, ending the brief era of nuclear expansion.

Nuclear power plants had their problems, but they did not pollute the air. Fossil-fuel plants did, emitting nitrogen oxide, a cause of acid rain, and carbon dioxide, considered the main contributors to global warming. (The Clean Air Act of 1970 took the first step in putting pressure on utilities to look at other sources of energy.) Then, in 1973, the Arab-dominated Organization of the Petroleum Exporting Countries (OPEC) placed a six-month embargo on oil shipments as a result of the United States' support for Israel during the Yom Kippur War. Problems developed in Iran, resulting in the Shah being deposed and American hostages being held. Oil prices rose and by 1980 reached almost one hundred dollars a barrel in today's dollars. The cost of oil, the dependence on its availability from foreign sources, and growing environmental concerns became major issues in the United States—issues that remain unresolved forty years later.

Congress took a step at the time to reduce the effect of electric energy production on the environment by funding wind energy research over the period from 1973 to 1988. Emphasis was placed on the development of bigger wind machines, the idea being that these machines would be more efficient and reduce the cost of wind energy. This led to companies such as General Electric, United Technologies, and Boeing working with NASA and the Department of Energy to build prototype wind turbines ranging in capacity from 2 to 4 megawatts. All had troubles—a storm destroyed a unit, a blade ripped off, a generator tore apart, and general reliability problems shut down units far too frequently. "Though well funded," wrote wind turbine authority Paul Gipe, "these massive turbines seldom performed as expected or operated very long before being scrapped, their builders moving on to other, more lucrative ventures."[6]

The question was why wind generators, which had been used success-fully years earlier, had so many problems in the 1980s. Wind energy histo-rian Robert Righter wrote, "Neither the efforts of private companies nor the infusion of federal dollars could instantaneously overcome the consequences of thirty years of neglect."[7] Of all the companies that participated in this de-velopment, only General Electric was manufacturing wind turbines twenty years later.

While the United States had expended vast sums with huge corporations to develop high-tech, large-capacity wind turbines, Denmark took a different approach. Craftsmen in small companies concentrated on "simplicity and very heavy construction," which resulted in smaller capacity, highly reliable units. The Danish government had encouraged this development by man-dating that utilities use wind energy and by giving investment tax credits and exemptions from energy taxes for wind turbine manufacturers. Vestas became the best-known wind turbines in the business.

CALIFORNIA BOOM AND BUST

Then came the "Great California Wind Rush." Several factors triggered the boom. First, smog and pollution became major issues in the state. Second, a "green" politician, Jerry Brown ("Moonbeam," as he was called), became gov-ernor and promoted conservation and investment in renewable energy. And third, California had three locations with geographical features that pro-vided outstanding wind characteristics: San Gorgonio Pass near Palm Springs, Tehachapi Pass east of Bakersfield, and the Altamont Hills Pass east of San Francisco. Each site contained a mountain pass through which cool air funneled eastward to replace hot air rising from the desert.

Meanwhile, in 1978, during the energy crisis, Congress passed the Public Utility Regulatory Policies Act (PURPA), the intent of which was to reduce the dependence on foreign oil, further encourage the use of alternative energy, and diversify the electric power industry. Utilities were required to use hydro, solar, and wind sources if the cost of production of these renewable sources did not exceed something called the utilities' "avoided cost." (Avoided cost is the cost the utility would have incurred had it supplied the power itself or obtained it from another source. It is the price at which an electric utility purchases the output of a qualifying facility, such as a wind farm.) This some-what nebulous figure took into account the cost (fuel and capital investment) supposedly avoided by the utilities in using renewable energy. Because of the

A Thousand Years

situation in the Middle East, the price of the fuel component of that cost (fuel oil, natural gas, and even coal) was abnormally high at the time, thus giving wind energy an advantage. Developers signed ten-year, fixed-price power contracts with the utilities. After that period the contract price would fluctuate with the market and eventually fall. On top of this, the federal government offered to renewable resource developers investment tax credits of 25 percent, accelerated depreciation, and low-interest small business energy loans.

California, adding to the federal government's benefits, ignited a wind energy boom by establishing a generous rate at which utilities had to pay for wind-generated electricity (avoided cost) of seven cents per kilowatt-hour. The state also provided a 25 percent investment tax credit plus tax-free state bonds to developers. Both the state and the federal government's tax credits applied to what the developer invested, not what was produced. As William Chapman, one of the pioneer wind developers, said, the people who designed the tax credits "had their heads screwed on backwards. It was all about how much concrete was in the ground, not how many kilowatt hours were generated."[8] In 1983 *Forbes* magazine called the situation the "investment fad of the year: the wind park tax shelter." Even the legitimate investment community was participating. Merrill Lynch sold wind partnerships to its clients, proceeding with three back-to-back investments worth $100 million each for wind farms in Altamont Pass.[9] For the investors, the tax incentives began in 1981 and lasted for four glorious years.

By 1985 over twelve thousand wind turbines were in place at the San Gorgonio, Tehachapi, and Altamont sites. With the failure of NASA and the Department of Energy to develop huge wind turbines in the 1970s, most of the units installed in California were small—in the 50- to 100-kilowatt range. Overall capacity for these thousands of units by 1988 reached almost 1,400 megawatts, 90 percent of the global total. But each of these sites had problems. At San Gorgonio, wind sandblasted the turbines, fouling up the rotors and scarring the blades. The Altamont winds "ripped to shreds" the blades of the early turbines. Blades hung from the towers like "crumpled birds."[10] And bugs built up on the blades, reducing efficiency. Also at Altamont, with sixty-five hundred units covering eighty square miles, the turbines killed thirty-nine golden eagles per year. Cynthia Struzik with the U.S. Fish and Wildlife Service wanted to shut down the site for illegal bird kills. In her mind the acceptable rate of bird kills was zero. At Tehachapi, ice and snow froze the turbine parts, including the rotors. But despite the troubles, a report for the California Energy Commission stated this was the beginning of the modern wind energy industry and the birth of the wind farm concept.

A Thousand Years

Wind turbines cover the hills in the Tehachapi Pass area of California. Wind farm development began in the state in the 1980s. (Courtesy of National Renewable Energy Laboratory.)

The wind energy boom did not achieve the historical significance of the California Gold Rush of one hundred years earlier. But like all booms, it soon became a bust. With the Reagan administration in power, federal tax incentives ended in 1985 and growth dwindled. As the situation in the Middle East stabilized, oil prices fell, making wind energy expensive for the utilities. With companies rushing to build wind turbines, failures occurred, keeping many of the units shut down. There were lawsuits and bankruptcies. One of the responsible developers, Bill Adams, described the situation: "A lot of schlocks were getting into the business and selling prototype machines on a mass basis that didn't work."[11] During an investigation of one site, Internal Revenue Service agents found employees attaching helicopter blades to units that hadn't worked in months.

Some might compare the California wind energy boom to another energy boondoggle, the synfuels tax credit. It too had its origins with the gas crisis of the 1970s. The theory was to encourage the conversion of coal to gas, but it evolved into something altogether different: If the chemical composition of coal was modified, it qualified for tax credits. Thus coal was sprayed with diesel fuel, pine tar resin, or other substances. Synfuel plants sprung up in the coal regions of West Virginia, Virginia, and Alabama. Investors, as in the California wind energy boom, jumped in, but synfuel was so attractive that a retail chain store and Marriott Hotels also participated. The synfuel coal was sold to utilities just like regular coal, and if it was sold at a loss, it didn't matter. By the synfuel program's expiration in 2007, it was estimated that investors had reaped billions in tax credits from the federal government.

While tax credits encouraged the proliferation of wind turbines, the view of thousands of these machines, particularly in the elegant Palm Springs area, raised opposition. One wind farm just off I-10, about ten miles east of Palm Springs, was described as "an eyesore of broken and twisted blades." When Hollywood notable Sonny Bono became mayor, he opposed further expansion. A city councilman declared that the turbines were "as damaging to Palm Springs visually as strip mining has been to towns and villages in Kentucky and West Virginia."[12] People living within two miles of the turbines complained of the "whooshing" noise, particularly at night, when the sound carried. But by 1990 the sentiment had begun to change. Mayor Bono became one of the converts. Palm Springs needed more tax revenue, and annexing wind farm sites into the city was better than raising local property taxes. Plus the community began to realize that the wind turbines were a tourist attraction, with cars and buses parked along the highways enjoying the sight. Nevertheless, well-organized local opposition continued. There were the "not in my backyard," or NIMBY, syndrome and concerns about bird flyways. An environmental columnist complained that the affluent population was "more concerned with property values than with ecological sustainability."

As Peter Asmus wrote, "The bust was culling inferior machines and corrupt opportunists."[13] The "schlocks" began to disappear as more reliable Danish wind turbines were installed and more responsible wind energy companies, such as U.S. Windpower (later Kenetech), SeaWest, and Zond, survived, at least initially. Kenetech, as both a developer and manufacturer, went bankrupt in 1996. SeaWest lasted as a wind farm operator and service provider to the industry, while Zond changed ownership a couple of times.

Founded in 1980, Zond succeeded by using the "more sturdy" Danish wind turbines (Vestas) until it built its own units in the 1990s. Its new 1.5-megawatt unit was designed in conjunction with a German manufacturer it had acquired. It soon surpassed Kenetech as the United States' number-one wind company. In 1997 Enron acquired Zond and provided the needed financial resources. When Enron imploded and went bankrupt, General Electric bought its wind turbine business in 2004 and formed GE Wind Energy, at present the primary manufacturer of large wind systems in this country.

And while federal government benefits for wind energy had expired in 1985, the Energy Policy Act of 1992 established the basis for a more substantial, less speculative growth. The act created the Exempt Wholesale Generator, or EWG. Public utilities now bought electricity from the cheapest source they could find, no longer depending on just their own generating units. As a result, independently owned natural gas power plants became the immediate beneficiary, but wind energy would become the growth industry.

The 1992 act also opened transmission lines, formerly controlled by each utility, to all comers at the same cost. Federal production tax credits and accelerated depreciation plus mandates by states stimulated the development of renewable resources. It was not until 1998 that the effects of the Energy Policy Act of 1992 began to renew the growth of wind energy, not just in California, but across the nation. And even then the growth was spotty, with down years in 2000, 2002, and 2004 because Congress was renewing the production tax credit on a year-to-year basis and in some years not renewing it at all. With the Energy Policy Act of 2005, Congress brought back the production tax credit, first for one more year, then extending it further. As a result, wind energy began to grow at an accelerated rate.

By 2000 entrepreneurs such as Invenergy's Michael Polsky had formed new companies (exempt wholesale generators) to develop both conventional power plants and wind farms like the Beech Ridge facility in West Virginia. These developers took advantage of this new competitive power market. Public utility Florida Power and Light's parent, NextEra Energy, formed a separate subsidiary and entered the market. A less scrupulous Enron, before its collapse, participated as well.

In California, the small turbines with capacities of 65 kilowatts (0.065 megawatts) still existed with their lattice-type, eighty-foot-tall towers. But as the twenty-first century began, they were disappearing as the newly developed three-hundred-foot, 1.0-megawatt units with narrow tubular towers took their place. One of these new units could replace six to twelve of the older types, and even bigger 1.5-megawatt units were on the way. The new turbines, with lightweight, composite, three-blade propellers could sense and respond to wind direction and wind-speed surges. New blade profiles captured more energy, thus increasing turbine capacity. There was hope that

23

A Thousand Years

these advancements could bring down the generation cost to five cents per kilowatt-hour, making wind competitive with fossil fuels.[14]

California wind energy capacity decreased from its 1980s peak, but the new, bigger turbines put California back in the growth mode and made it the leader, along with Texas, in providing wind energy in the United States. While the growth rate was impressive, wind energy expansion in 2006 equaled an amount equivalent to only two large coal-fueled power plants, consequently generating in total less than 1 percent of the electricity in the country.

An impetus for further expansion of wind energy had come in 1997, but more for Europe than the United States. The industrialized nations of the world met in Kyoto, Japan, to draw up an agreement to reduce greenhouse gas emissions as a means to combat global warming. These emissions, particularly carbon dioxide, were to be reduced by 2012 to 5 percent below the level emitted in 1990. To come into force, the treaty had to be ratified by the nations emitting over half of the worldwide emissions. The treaty took seven years to ratify and became effective in 2005. Different countries agreed to different levels of emission reduction. The United States was the world's largest emitter at the time; it refused to sign. President George W. Bush declared that without countries such as China and India participating, it would not be economically sound for the United States to participate.

On the other hand, the European Union agreed initially to an 8 percent emissions reduction and charged ahead to grab world leadership in wind energy.

Chapter 3

Europe: Reality or Tilting at Windmills

WORLD LEADER

California may have gotten the jump on wind energy development in the 1980s, but Europe, with its history of hundreds of years of windmill usage, quickly took the lead in the 1990s. By the beginning of 2012, Europe had 40 percent of the world's installed wind energy capacity. This compared to 21 percent for the United States and 33 percent for the surging Asian countries of China and India (see Table 3.1). Germany led the European countries, followed by Spain.

Though it fell out of the top ten in the world after being ranked there for many years, the small country of Denmark tended to get the most publicity regarding wind energy, with presumably 20 percent of its power generated by wind. Europe, led by the United Kingdom, was also the world leader in offshore wind farms, with 3,000 megawatts of installed capacity.

Between 2009 and 2011 there had been a dramatic shuffle in world wind energy leaders. Germany led in 2009; the United States moved ahead in 2010. But China far exceeded its declining-growth rivals in 2011, adding 37,000 megawatts of capacity in just two years, a number that exceeded Germany's twenty-two-year total.

Unlike the United States, European countries established national goals for the use of renewable resources. And, unlike the United States, they embraced the Kyoto Protocol, the intent of which was to reduce greenhouse gas emissions. Meanwhile, China, under authoritarian rule, has developed the world's most extensive wind energy industry.

While leading the way in wind energy development, European countries have experienced problems, many of which have been touted by opponents of wind energy expansion in the United States. Opposition organizations, such as Mountain Communities for Responsible Energy, which fought the Greenbrier County project in West Virginia, have used these problems to question and in some cases to slow down wind energy expansion.

DENMARK'S 20 PERCENT

Compared to other countries, Denmark already had a long wind energy history. While Charles Brush was operating his wind turbine in the United States in the 1880s, the Danes were adapting windmills to generate electricity. By 1906, Denmark had forty windmills producing electric power, thus setting a standard that continues today.

Table 3.1. Top Ten Countries in Total Wind Capacity, December 2012		
Country	Capacity (MW)	Percentage of Total
China	75,564	26.8
United States	60,007	21.2
Germany	31,332	11.1
Spain	22,796	8.1
India	18,421	6.5
United Kingdom	8,445	3.0
Italy	8,144	2.9
France	7,196	2.5
Canada	6,200	2.2
Portugal	4,525	1.6
Total Top Ten	242,630	85.9
Other countries	39,853	14.1
World total	282,483	100.0

Source: Global Wind Statistics—2012, Global Wind Energy Council, February 2, 2012.

Europe: Reality or Tilting at Windmills

A traditional windmill and modern wind turbines coexist in the Dutch countryside.
(Photograph by Paul Van Beets, Dreamstime.com.)

The 1973 oil embargo, initiated by Arab oil-producing states, provided further impetus. Denmark restricted energy usage. People took cold showers and limited their driving. The government decided to no longer depend on energy imports and to find renewable resources. Wind was that resource, and the government subsidized wind energy development. As early as 2003, the country had five thousand turbines installed both onshore and offshore and had the highest percentage of total power capacity devoted to wind of any other country. Meanwhile, its wind turbine manufacturing capability captured a third of the world market. In the 1980s, Danish manufacturers even began exporting wind turbines to the United States.

An *NBC Evening News* broadcast in 2007 praised Denmark, stating that wind turbines supplied 20 percent of the country's electricity "needs."[1] This figure was a popular misconception, as wind energy opponents like to point out. In reality, Denmark exported much of the wind energy it generated.

The NBC program described the Danish island of Samso, with its offshore wind turbines, as energy independent. In its own way this twenty-mile-long island, with a population of four thousand, is a microcosm of Denmark itself. Writer S. R. Nunnally described the island's offshore wind farm as being "composed of 10 beautiful turbines" and noted that this "Eco-Revolution" island received 100 percent of its electricity from wind power. What she did not mention was the obvious: When the wind doesn't blow, the island must receive electricity from the mainland. On the other hand, when the output from the wind farm exceeds the island's need—as at night, when demand is low—electricity must be offloaded. The excess is exported to the mainland's regional grid.

Like NBC and other media, *U.S. News &World Report* also made the claim that in Denmark wind power provides 20 percent of the nation's electrical needs.[2] The article, however, recognized that because of wind's "flighty nature," the Danish government had to buy electricity from the hydro- or fossil fuel–powered plants of its neighbors while at times selling its overproduction cheaply or at a loss. In addition, wind energy subsidies were a factor in making Denmark's tax burden "among the heaviest in the world," and its electrical rates were the highest in Europe.

Before writing for the *Country Guardian*, the publication arm of an organization in the United Kingdom concerned about the environmental and social aspects of commercial wind farms, V. C. Mason researched and evaluated the situation in Denmark for several years.[3] With its many turbines and a population of five million, the country has the highest ratio of wind turbines

Europe: Reality or Tilting at Windmills

to population in the world, and its turbines are sited on a basically flat, agricultural landscape.

The Danish wind industry's impact on reducing carbon emissions, according to Mason, was negligible. The energy going to Norway and Sweden supplemented "green" hydro-electricity production, not fossil-fuel energy, and thus had no effect on emissions. And there had been "little or no permanent closure" of fossil-fuel plants in Denmark. Backup generators had to be retained to "shadow" the output of wind power and balance load demands on the domestic grid. Online spinning reserves, operating at low efficiency, spewed greater amounts of carbon dioxide per kilowatt-hour. Over the years, roughly 70 percent of Denmark's electricity generation has been powered by coal and natural gas, and there has been no significant change. The country has no nuclear power.

A study by Denmark's Center for Politiske Studier (CEPOS) declared that Denmark's "highly intermittent" wind power during a five-year period provided an average of 9.7 percent of the country's annual electricity consumption. In one year it reached a low of 5 percent.[4] The country's excess power, that unable to be integrated into the country's grid system, is offloaded at a loss to Norway and Sweden. These countries rely heavily on hydropower and can absorb Denmark's fluctuating wind energy surplus by rapidly cutting back water flow and thus reducing power output. The electricity when available from Denmark can also be used to pump water to elevated reservoirs that feed the hydro plants.

Mason further stated, "With a better understanding of the technical and environmental limitations of wind technology, public opposition has grown." There is a concern that wind turbines have "seriously detracted from the charm, beauty and peace of Danish landscapes and coastlines." And while the government had put pressure on local municipalities to find sites for larger turbines, protests all but stopped the erection of these units.

Despite overload problems and the recent lack of growth, the Danish government had established an objective of meeting 50 percent of the country's electricity consumption with wind by 2025. This compared with the European Union's objective of 20 percent by 2020. With this objective, Denmark would double wind power capacity from 3,000 to 6,000 megawatts. In recent years, however, there has been little growth in Danish wind energy. It took Denmark's capacity three years to grow from 2,000 megawatts to 3,000 megawatts, but nine years to reach 4,000 megawatts in 2012. More growth demanded the decommissioning of its older and smaller units, replacing them with new, larger turbines. More onshore and offshore sites needed to be designated and the cost of offshore units reduced. Integration problems had to be solved by closely coordinating and improving interlocking grids between Denmark and Germany, Norway, and Sweden. Government subsidies were to continue for the twenty-year life of the turbines. Subsidies supplemented Denmark's market price for electricity, with a production tax credit that would be increased further if a wind developer replaced old turbines with new ones. The CEPOS study indicated that Denmark's surplus would increase in 2013 when 800 megawatts of new offshore capacity is commissioned. However, nearly all of this would be exported again "without achieving any significant fossil fuel or any $CO2$ reduction."

Energinet, the independent, public enterprise that owned and managed the electricity grids, reported, "If half of Denmark's power consumption is to be covered by wind turbines, Denmark must develop a far more intelligent and flexible power system."[5] Denmark had significant work to do to put wind energy back on a growth path to meet its own long-term goals and contribute to the European Union's overall objectives.

GERMANY'S SEARCH FOR REALITY

Though Germany got its start ten years after the California wind boom began, it quickly became the biggest generator of wind power in the world. The Chernobyl disaster in 1986, government legislation in 1990 and 2000, and the signing of the Kyoto Protocol drove its growth in wind energy. In 1997 it surpassed the United States, primarily California, in wind-generated installed capacity, and over the next ten years, its wind capacity grew tenfold, reaching 28 percent of the world's total. However, by the end of 2012, Germany's growth rate had dropped significantly and represented only 11 percent of the world's total. The United States had surpassed it in total capacity by 2008, and China matched it a year later.

Thirty-seven of Germany's twenty thousand wind turbines were located on hills overlooking the small farm town of Alsleben. This rural scene in eastern Germany looked much like the Beech Ridge Wind Farm in Greenbrier County, West Virginia, was expected to look when completed. This wind park, as it is more commonly called in Europe, was owned by General Electric and used GE turbines to provide 54 megawatts of installed capacity. It was one of the largest wind parks in the country at the time, an indication that Germany consisted of a great many small projects scattered across the countryside.

With a strong "green" movement in Germany, the government had plans to phase out nuclear plants and to compensate the loss with wind energy. The country established an objective of 20 percent of its energy coming from wind by 2020. To accomplish this, at least $1.4 billion in transmission lines had to be added. Despite the political popularity of reducing greenhouse gases, the grid expansion cost caused concern due to Germany's already high cost of electricity. The industrial sector might not be able to absorb these costs and stay competitive, plus domestic consumer prices would be driven higher.

Grid expansion was not the only problem Germany faced with wind energy. Affected communities were objecting to the wind turbines, plus the government was lowering fixed-price subsidies for the industry. The German Wind Energy Association acknowledged that without subsidies, "the wind energy sector would have no chance against the billion euro heavyweights of the coal and atomic energy industries or the cartel-organized energy market."[6] The tariff was going to decrease at a set rate per year, but the association believed (hoped) that by 2015 the cost of wind energy would decrease to the point that it would be cheaper than fossil-fuel power.

Meanwhile, the installation of new turbines was decreasing. While export of wind turbines by Germany's wind equipment manufacturers remained favorable, the concentration at home was on replacing old turbines with more efficient, taller units. Hermann Albers, president of the German Wind Energy Association, expressed concern "that the fundamental conditions for wind power utilization are no longer favorable in Germany." Local citizens raised objections about the countryside being spoiled and tourism being impacted and were concerned about the same turbine noise factors that wind developers in the United States were encountering. German courts heard several hundred cases a year related to local objections to wind projects. A court in Darmstadt, Germany, ruled against two turbines being placed in the city, stating, "They would alter the character of the local environment." Offshore wind parks, which faced fewer objections, were considered a means of expanding Germany's wind energy, but they faced regulatory and technical problems.[7]

The E.ON Netz company operated a grid supplying electricity to twenty million people located in an area stretching through the middle of the country from the North Sea to Austria. It assimilated almost half of Germanys' wind power into its grid and probably had more practical experience in integrating large amounts of wind power into the system than any other operator. The company shared that experience in its "Wind 2005" report and supplementary 2006 report, taking a frank approach in presenting data about

Europe: Reality or Tilting at Windmills

wind energy's shortcomings. The company was committed to wind energy but felt the keys to success were to make renewable energy competitive and to improve its integration into the electricity grid.

If Germany wanted to double its wind power capacity by 2020, there were three challenges, according to an E.ON Netz. One: "Wind energy is only able to replace traditional power stations to a limited extent." Two: "Wind power feed-in can only be forecast to a limited extent." And three: "Wind power needs a grid infrastructure." The E.ON Netz report concluded that the "possibility of integrating this generation capacity into the supply system remains to be seen."[8] One year with 7,600 megawatts of wind power installed, the feed into the grid ranged unevenly by month from a high of 6,234 megawatts available in January to only 8 megawatts in September. The average capacity factor for that year was 18 percent.

During Christmas week in 2005, wind power feed-in during a ten-hour period fell from 6,000 megawatts to 2,000. As the E.ON report stated, this 4,000-megawatt drop had to be backed up by the equivalent of eight 500-megawatt coal-fired power plants. As wind power capacity increases, the variability increases and there is a further impact on the reliability of the grid. As a result, the study contended that if the 2020 goal of 48,000 megawatts of capacity is achieved, only 2,000 megawatts could be considered reliable. (Those who opposed wind energy frequently referenced the E.On Netz report)

The German Wind Energy Association's Christian Kjaer rejected these grid-related objections, saying, "All electricity grids are designed to cope with power fluctuations." He argued, in somewhat of a fanciful way, that "fossil fuel or nuclear power stations are truly intermittent. You never see 1,000 megawatts of wind energy shutting down in a second, yet that's what conventional power stations do."[9]

By 2009 Germany's wind energy goals had been increased to meet 25 percent of the nation's needs by 2020, despite the decreasing growth in capacity. This new objective now required 55,000 megawatts of installed capacity compared to the 29,000 at the beginning of 2012. The German Federal Transport Ministry declared ambitious goals for offshore wind energy, announcing forty wind farms, up to 124 miles off the coast, totaling twenty-five hundred wind turbines with 12,000 megawatts of installed capacity. With this expansion, Germany would obtain 12 percent of its electricity consumption from wind, double the current amount.[10] However, by the beginning of 2012, Germany had installed only fifty-two turbines with a

Europe: Reality or Tilting at Windmills

capacity of 200 megawatts offshore. In the following six months, twenty-seven more turbines were installed.[11]

The country's national climate protection targets were "envisioned" as being 30 percent consumption from all renewable resources by 2030. But again there were concerns: strict environmental regulations, the cost associated with placing and maintaining turbines at sea in 130 feet of water and nineteen miles from shore to protect wetlands and the tourist industry, the high investment risk of wind versus nuclear and coal-fired plants, and the ever-present grid integration problems.[12] At the higher capacity goal, generation could be higher than needed when consumption was low. In some grid areas, this had already happened. Like Denmark, a need existed in Germany for an overall more flexible power system and for storage, such as integration with hydro plants.

With Japan's 2011 Fukushima nuclear disaster, Germany further complicated its energy situation by immediately shutting down half of its nuclear plants. This was followed by a decision to phase out the remaining plants by 2023. Nuclear power provided 23 percent of Germany's electricity. This move by Chancellor Angela Merkel's government was characterized as an emotional, not technical decision.

To replace the power from the nuclear plants, Germany looked to the expansion of renewables, especially offshore wind energy, but faced the reality of importing more coal, natural gas, and nuclear power from its neighbors. With this came an increased cost to consumers and industries and the increased difficulty in meeting its CO2 reduction goals. The costly need for upgrading its transmission system, an effort that could take ten to fifteen years, became an even greater necessity.

With a frankness unusual among wind energy association leaders, Hermann Albers, president of the German Wind Energy Association, said, "The government is trying to sell the expansion of offshore wind energy as a miracle solution for its nuclear phase-out." He noted that these projects were "massively" behind schedule. The financing was risky and complex, and the technology was still a few years away from delivering significant amounts of electricity. Albers believed onshore projects would offer more rapid and more efficient results and that there were still some regions, particularly in southern Germany, available for development. The government, already required to buy wind power first, irrespective of competitive price, had increased the feed-in tariff for offshore wind farms to twenty-five cents per kilowatt-hour, more than triple the rate for energy on the spot market.[13]

Europe: Reality or Tilting at Windmills

With its nuclear phase-out and its wind energy growth leveling off, the German government continued to raise renewable energy expectations. Where was the reality?

SPAIN: OVERDOING A GOOD THING

In Spain, windmills have been a part of the culture since Don Quixote tilted with them in 1600. However, the country's introduction to wind energy did not begin until 1994, when the government introduced production subsidies. By 1999 Spain had become one of the first European Union countries to establish a specific renewable energy target. The 2010 target was surpassed six years early. Spain was matching Germany's early 30 percent growth rate. Wind power was seen as a solution to the country's heavy dependence on importing natural gas. With strong government support and a long coastline, Spain faced fewer restrictions on wind energy expansion than Germany and was projected to become Europe's future leader.

Spain established ambitious goals for 2020, hoping to eventually have renewable energy supplying 48 percent of the country's power needs. To achieve this goal, wind predictions had to be improved to assure that wind fluctuations could be handled at high levels of penetration to the grid. Transmission operators, utilities, wind developers, and regional governments had to develop a "strategic grid framework" to provide centralized control of the grid. The installation cost of turbines in Spain was decreasing. From 1986 to 2006, the cost to install a megawatt of capacity had dropped from $2.2 million to $1.3 million.[14] Like Denmark and Germany, Spain manufactured turbines and their components for both domestic and international markets.

As part of its 2020 goal, the government was looking at offshore wind locations. Spain, a peninsula with sea on three sides, was in an excellent geographical position, and a number of offshore zones were established. The government required developers to demonstrate that their wind farms would generate at least 50 megawatts of electricity and would not ruin the environment. Meanwhile, questions about the impact on tourism were limiting wind farm development on the mainland as well as offshore. A shallow area off Spain's southern Atlantic coast where five hundred turbines were to be placed generated a concern. At the scene of the Battle of Trafalgar, opponents raised objections about desecrating a historical site. By 2012 Spain had negligible offshore wind capacity.

Europe: Reality or Tilting at Windmills

In addition to designating locations where tens of thousands of turbines could be placed, Spain intended to increase the use of hydropower and add solar capacity. However, sufficient natural gas capability would remain as backup when the wind didn't blow. Nuclear power would continue as a backbone to the system. Coal was to be phased out. The Spanish government, which the *London Telegraph* described as "socialist," had established not just a goal but also a detailed plan to achieve it, something the government of the United States' democratic process had been unable to achieve.

In 2009 Spain led Europe in wind energy, with an increase of 2,459 megawatts in installed capacity. Wind provided an increasing percentage of Spain's electricity demand, reaching 14 percent and exceeding coal's 13 percent for the first time. Natural gas provided 30 percent and nuclear power 20 percent. On the morning of December 30, 2009, wind energy supplied 54 percent of demand as a result of high winds. These results indicated the success Spain had achieved in upgrading its grid with the ability to integrate large amounts of wind into its power system.

But despite the growth in wind energy, Spain's economy was in trouble. The *New York Times* reported the country had a "huge" deficit and high unemployment. The governor of the Bank of Spain said that without reforms, the economy faced a "tough and complicated period." The government appeared "to bumble ahead confusedly, casting proposed cutbacks into the air, then reeling them back in."[15] Both solar and wind energy were affected.

Spain's Congress of Deputies approved a royal law decree replacing a 2007 decree that had been favorable to wind and solar energy growth. The new law controlled the rate of installation by limiting the system of premiums and tariffs previously established. The president of the Spanish Wind Energy Association called the situation "very worrying." His organization stated that newly installed capacity would be limited to less than 800 megawatts for each of the next three years.[16] The government policy of promoting wind and solar energy had caused a "rush" into the industry. Lavish subsidies had resulted in inefficient solar sites and increased costs. A new renewable energy plan had to be approved to ensure that Spain would meet the European objective of 20 percent renewables by 2020.

Spain had become an example of over-enthusiasm for wind energy. The flip-flops in government policy had initiated the subsidy flow and then reduced it. This was similar to the situation in the United States that encouraged then ended the 1980s California wind boom. In the 1990s and 2000s, off-and-on production tax credits slowed wind energy development.

Wind energy, not yet ready to stand on its own feet, was dependent on government, not the market economy.

Spain added 1,000 megawatts of capacity in 2011, an annual increase of 5 percent, the lowest rate in the history of its development. However, on the production side, wind covered 26 percent of demand in April 2012, matching its nuclear generation. Though down from 2010, wind covered 16 percent of demand for the year 2011. For one hour wind supplied 61 percent of the country's demand while operating at a 77 percent production capacity factor. In 2010, while having 18 percent less installed capacity than Germany, Spain produced 25 percent more electricity, indicating a striking difference in the two countries' wind systems.

Despite these strong generation numbers, there was some question whether Spain would add any wind capacity in 2013. Due to the countries' financial problems, a moratorium ended subsidies, the highest in Europe, for all new wind systems beginning January 1, 2012. In fact, there was talk of the government adding an 11 percent tax on wind power generation. This threatened the marginal profitability of existing wind operators, most of whom carried a heavy debt load. Spain's T-Solar and Iberdrola, both large wind farm operators, and Gamsea, the Worlds' fourth largest wind turbine manufacturer, were moving their operations to other countries in order to stay in business. Employment in the wind industry had dropped to thirty thousand, a decrease of ten thousand with more to come. This was occurring in a country where total unemployment was already 25 percent and indebtedness was threatening the stability of the euro. In addition, Spain's utilities were running at a deficit due to the government's establishing consumer utility prices that were insufficient to cover increasing power costs. Spain had achieved a high level of wind energy generation, but at a cost that was unsustainable.

UNITED KINGDOM: THEY HAVE TO GO SOMEWHERE

Despite having great wind resources, maintaining consistent growth, and leading Europe in offshore wind capacity, the United Kingdom remained far behind Germany and Spain. Wind did represent 5 percent of the country's source of energy; however, the government's goal was once 30 percent by 2020, an increase that even the British Wind Energy Association described as "masssive." What seemed more reasonable was the European Union goal of 15 percent by 2020. But even then, one critic said it would take fifty thousand turbines covering an area the size of Wales to make this goal. Onshore wind farms were mostly small, spread across the highlands of Scotland and Wales. However, one of the largest wind farms in Europe, the Whitelee Wind-farm outside Glasgow, Scotland, covered twenty square miles and had 144 turbines providing an installed capacity of 322 megawatts. According to the European Wind Energy Association's wind map, Scotland, surrounded by the sea and with mountain heights to four thousand feet (a height similar to West Virginia's ridgelines), had the best wind potential both onshore and offshore of any place in Europe.

But the barriers to wind energy were tough to knock down. In the past the government had rejected a twenty-seven-turbine wind farm, the largest proposed in England at that time, to be placed between the Yorkshire Dales and the Lake District. The tourist board opposed it and a naturalist swore he would chain himself to a turbine if construction went ahead. At one point there were 150 anti–wind farm organizations in the country, all objecting to the conversion of rural landscapes into industrial landscapes. The Campaign for Rural England encouraged placing wind farms offshore. The anti-wind groups raised the usual arguments in addition to viewshed—the noise, the threat to birds, the expense—and argued that energy conservation was more effective.

The British government changed its renewable strategy plan to more easily put turbines onshore. To reduce carbon emissions by 34 percent by 2020 would require 26,000 megawatts of capacity, totaling ten thousand turbines with four thousand offshore. There were then twenty-five hundred on land. To the critics of wind turbine size, landscape disfigurement, and noise, Energy Secretary Ed Miliband said government ministers would be sensitive to residents' concerns. But, he added, "they [the turbines] have to go somewhere."[17]

John Constable of the Renewable Energy Foundation, a think tank, called the government's renewable energy strategy "hugely expensive" and "wildly optimistic." Furthermore, he said, it would reduce "just seven percent of the UK's annual CO2 emissions and only 0.1 percent of world emissions." He was also concerned about what was not being considered. Over the next ten years, one-third of the country's generating capacity would be lost because of aging nuclear plants and coal-fired plants that would be shut down to meet European Union emission standards.[18] This situation was illustrated to some degree during Britain's cold snap in January 2010. With a cold front and a lack of wind, wind energy, with the potential of producing 5 percent of the

37

A ferry cruises past offshore wind turbines in the Irish Sea near the English coast. (Photograph by Steve Heap, Dreamstime.com.)

country's power, could only produce 0.2 percent. The director of the Energy Intensive Users Group, Jeremy Nicholson, warned that this could turn into a crisis: "If we had this 30 gigawatts (30,000 megawatts) of wind energy, it wouldn't have contributed anything of any significance this winter." Even Secretary Miliband was giving consideration to nuclear power as a means of filling the energy gap.[19]

With the public resistance to onshore wind, and data that indicated that the capacity factor in 2008 had been 29 percent for onshore operations and

Europe: Reality or Tilting at Windmills

35 percent offshore, the future for British wind energy lay in the stronger winds offshore, despite the significantly higher cost of installation and operation and the higher subsidies.

The premier offshore wind farm became the London Array. Planning for the farm began in 2001, government approval came in 2007, and construction started at the beginning of 2012. By the end of that year, Phase I was in place, with 175 turbines with a capacity of 630 megawatts. When Phase II is completed, the wind farm will cover 152 square miles and provide 1,000 megawatts of capacity.

The Array has had problems. A consortium of three companies planned to invest what was once 1.5 billion euros (over $2 billion in U.S. currency) in the project, but the cost grew to 2 billion euros. At that time, the *London Telegraph* reported the Array was "said to be on 'knife edge' because of rising costs." In 2009 Royal Dutch Shell dropped out of the consortium, energy prices fell, and loans became difficult to obtain, but three developers pushed forward: E.ON, the German utility that had worried about absorbing large amounts of wind energy into the German grid; Dong Energy, a Danish company; and Masdar, an Abu Dhabi firm. The group declared the London Array financially viable, crediting the government's increased support through its renewables obligation certificate scheme, which requires British utilities to have 12 percent of their sales in 2012/2013 from renewable energy or pay a penalty (government subsidies were weighted two to one in favor of offshore wind versus onshore). Of course, electricity customers paid for the scheme.[20] E.ON and Dong apparently saw more potential for wind energy expansion in Britain than in their home countries.

The Royal Society for Preserving Birds also raised objections to the project because of concern for the red-throated diver, supposedly "a bird rarely seen in UK waters." However, seven thousand of these birds had been counted in the area between 2002 and 2005. As a result, the developers agreed to reduce the number of turbines in the first phase. A borough council also had an objection to the project's onshore substation in their community. Supporting the London Array, the Friends of the Earth, an international group campaigning for solutions to environmental problems, noted that 1 percent of electricity used in the UK would come from the project. One of the organization's spokesman, Martin Williams, said, "It's really a big wind farm, but when you look at the scale of the challenge of climate change, we're going to need a lot more than just one more offshore wind farm."[21]

Europe: Reality or Tilting at Windmills

Great Britain was trying its best as the world's leader in offshore wind farms. By the end of 2012 it had an installed capacity of 3,000 megawatts offshore, and in the following years it expected to have as much capacity offshore as onshore.[22]

Yet problems remained. For onshore wind farms, Parliament cut subsidies 10 percent in 2012 and had considered 25 percent. The instability of government policy, with leadership lacking as pro-wind fought pro-nuclear interests and as politicians voted based on keeping their seats, affected the investment climate. Public concerns continued over the cost of subsidies, rising electricity prices, and objections to the view of onshore turbines.

GAMING CAP AND TRADE

Even with its strong growth in wind energy, Europe early on was not yet meeting its Kyoto Protocol carbon emission goal of 8 percent reduction by 2012. From 1997 to 2007, the European Union had reached a reduction of only 2 percent. Of the three major wind-producing countries, Germany was almost meeting its goal, Denmark was well behind, and Spain, despite its aggressive wind policy, had actually increased emissions. Spain's increase came from higher fossil-fuel electricity production, while low water levels caused a reduction in hydropower output. But that would all change for the wrong reasons.

Europe's process for meeting emission goals was cap and trade. In this system the governing authority establishes a ceiling on emissions and issues a corresponding number of permits allowing each company to emit a certain amount of emissions up to the cap limit. Utilities and other polluting industries buy and sell permits to meet their caps. A utility burning coal could buy permits to meet its cap rather than converting some of its power sources to, for example, wind. A utility largely dependent on nonpolluting nuclear power might have extra permits that they would be willing to sell. The exchange takes place on a greenhouse market.

A cap and trade system in the United States had reduced sulfur dioxide and other pollutants causing acid rain. This came out of the 1992 Clean Air Act and was considered by some to be the most successful environmental policy Congress had ever enacted. But acid rain was one thing. The climate change problem, being worldwide, was bigger and far more complex.

The European Union's existing cap and trade system was criticized for letting companies "game the rules," and it had failed to meet goals to reduce

40

emissions. The World Wildlife Fund reported that while "the mechanism of carbon trading was sound in principal, the first phase of the European Union scheme had been seriously undermined by weak political decisions. The operation covered only 40 percent of the emissions, excluding such areas as aviation, and did not impose strict enough emission limits on countries."[23] Caps on emissions were so loose that companies and utilities could meet the caps without making the cuts. Regulators had given away an excess of pollution permits. Polluting industries, factories, and power plants sold their permits to enhance their balance sheets yet retained sufficient permits for the future. Steel and cement plants benefited the most. Carbon market prices needed to be raised, but this caused concerns about low-cost competition from outside the European Union. As the *New Republic* noted, the plan "looked like a total flop."[24]

However, in 2009 the European Environmental Agency projected that Europe would reduce emissions by 13 percent by 2012, thus exceeding the 8 percent goal, and was on the way to meeting its 20 percent goal by 2020. The turnaround was attributed to corrections to the system. The European Environmental Agency stated in essence that the higher reductions would be achieved by several means. Roughly 7 of the 13 percent would come from the cap and trade policy and renewable energy efforts already in effect. Four percent would come from new energy policies that were planned—better forest management (forests absorb carbon dioxide)—and credits from financing energy projects in underdeveloped nations. Buying excess credits from Kyoto countries would provide the final 2 percent.

A decline in emissions occurred when manufacturing operations left Europe for other areas, though some believe that emissions actually grew if one considers imports from China. Nevertheless, shifting manufacturing elsewhere could hardly count as reducing emissions since there was no net worldwide reduction. A real decline was occurring not because of cap and trade but because of the business decline of 2008–9 and its continuing effect on the European economy. By 2012 emissions were below the current cap.

Meanwhile, the price for a ton of carbon on the market had fallen from forty-seven dollars in 2008 to seven dollars. At the low range, the penalty for emissions was such that there was little incentive for polluting companies to buy permits or invest in carbon reduction measures. The practical solution was to lower the cap on emissions to tighten the market and drive up prices. The European Union considered a weaker approach, the setting aside of units, which when announced had little effect on prices.[25]

Europe: Reality or Tilting at Windmills

And while Europe had its problems with cap and trade, it at least had a goal and was trying to meet it. Meanwhile, in the United States, the green revolution got a late start. The country had no emissions reduction standard, and the politics didn't favor one. Despite this situation, wind energy, though lagging Europe, began to grow. A wind farm on the mountain ridges of West Virginia was typical of this growth and the challenges that had to be overcome.

Chapter 4

Battle Begins

HOSTILE ENVIRONMENT

On a July afternoon in 2005, Dave Groberg spoke to a group at a middle school in Rupert, Greenbrier County, West Virginia. "We have purposely situated the project in a rather remote location," he said, "and it has been planned in such a manner that it will have minimal impact on county residents."[1] The speaker was a trim, young fellow in a dark suit and tie who had introduced himself as the manager of business development for Invenergy, a Chicago, Illinois, company. County politicians, economic development leaders, and media representatives listened intently.

Dave Groberg described a project of more than one hundred wind turbines to be spread along twenty-three miles of ridgelines in the vicinity of nearby Cold Knob. Having mountain ridges with sufficient, available property, and the nearness to a transmission line to feed the electricity to a regional grid, made the Greenbrier County site ideal for a wind farm. This electricity would be sold to utilities in the mid-Atlantic region. The facility would be located on land owned by the MeadWestvaco Corporation, a paper company that had cut timber on the land for a century. Dave Groberg stated that an Invenergy subsidiary, Beech Ridge Energy, as the facility would be called, would apply to the state for permission to proceed in the fall. Construction could start in the summer of 2006, and the wind farm could be operating by December of that year.

Up to the time Invenergy announced its West Virginia wind farm, wind energy development in the United States had been bumping along with one-year renewals of the production tax credit. But with the 2005 energy act, it looked good for the next three years, giving developers the sense that there was time to begin construction of new wind farms.

From the audience at that Greenbrier County school came a question: Had Invenergy received concerns about the aesthetics of the huge wind turbines at its other operations in Idaho, Colorado, and Montana? Dave Groberg said yes, the "viewshed issues" had come up. He noted that the Public Service Commission of West Virginia (PSC) would take comments from the public before making a decision. The commission was the sole approval authority. Greenbrier County's local government had no say; there were no zoning requirements or height restrictions in the county.

Greenbrier County had a population of thirty-five thousand and a total of five stop lights, most of them in Lewisburg, a place recently discovered by the *New York Times* and described as "a country town with a cosmopolitan edge." Situated in a broad valley of cattle farms, this historic community had a downtown that included galleries, antique shops, restaurants, a live theater, a museum, and a charming inn. To the east was the Greenbrier River, and through a gap in the mountains was the famed Greenbrier Hotel in White Sulphur Springs. The western, more mountainous part of the county had been mined and timbered. It was there that Invenergy wanted to place a wind farm, but many of the turbines would overlook a valley.

The news of a giant wind farm to be placed on the tops of the surrounding ridges came as a surprise to the people of Greenbrier County, particularly to the residents of Williamsburg Valley, where the turbines would distract from the forested mountain skyline surrounding the area's farmland. Quickly a fledgling group called a meeting in the Williamsburg Community Center, a frame building butted up against the local volunteer fire and rescue station. Fifty people attended, listening to concerns and complaints about wind turbines in general and the Beech Ridge Wind Farm in particular. Williamsburg native Debbie Sizemore led the meeting. As the community health coordinator for West Virginia, she traveled the state assisting local health departments, but she found time to be active in just about everything to do with her community.

Debbie Sizemore had never felt strongly one way or the other about wind energy. She had even seen the wind turbines as she traveled through Tucker County in northern West Virginia. This was the only wind project in the state at the time. As she described it, she came over a hill on U.S. 219, saw a blade protruding into the sky, and immediately thought a jumbo jet had crashed.[2]

John Stroud, a rare book dealer who owned a farm nearby, spoke at the community center meeting. He had done some research. He said, "Wind

Battle Begins

Cattle feed in the Williamsburg Valley, West Virginia. Invenergy planned to place wind turbines on the surrounding ridges. (Courtesy of Mountain Communities for Responsible Energy.)

farms are not being built to solve an energy problem but for financial gain." He went on to explain that wind farms were "first rate tax havens, thanks to federal and state subsidies." Stroud cited studies that showed physical and psychological risks for residents near the installations. "The noise never stops," he said.[3]

The discussion led to other concerns. Some said the towers were unsightly and would ruin the view. The turbines, others said, threatened wildlife and the blades killed birds. Locals hunted on the ridges and would no longer be able to do so. Many were concerned that the natural beauty of the mountains would be ruined, tourism would decline, and historical sites would be desecrated. Property values would drop. The turbines really produced very little energy, it was noted, as they were dependent on the wind blowing. And the energy would not be used in West Virginia but would be transported by transmission lines to other states. The ten to twenty permanent jobs the wind farm would provide were hardly significant.

"They think we're an ignorant bunch of rednecks," said Debbie Sizemore. "We need to organize and turn out in force for the public meetings to be held in October."

And organize they did. With Sizemore and Stroud as co-chairpersons, Mountain Communities for Responsible Energy was established. The organization's website stated in a high-minded way that its goals were to compile and assess information regarding the environmental and economic impacts of energy projects in Greenbrier County and explore unaddressed problems with wind energy. It would promote conservation programs and clean, affordable energy projects that were responsibly developed. But in reality the new organization had only one purpose: to defeat the Beech Ridge wind project.

Invenergy set up an afternoon open house with displays at the Williamsburg Community Center. Debbie Sizemore announced a meeting for that evening. Invenergy and Mountain Communities met in the center before a packed house of two hundred people. Resident Alice Crowe described the gathering in a letter to the editor to a local paper:

> On the night of October 12, 2005. . . . Families with their old and young were present along with a few state dignitaries and people finished with a day's work on their farms or woodland. Emotions were running high as they contemplated the implications of a literally towering industrial complex on their lives. The mountains themselves, clothed in October colors, seemed to sit silently. But they weren't voiceless. The overwhelming beauty of their presence spoke louder than only words could.
>
> Mountains have a way of enfolding and sheltering their people, making them feel protected from the ravages and chaos of the wider world. Mountain people feel protective in turn, of the mountains, and that was why so many came to save them.[4]

Beech Ridge Development Manager Dave Groberg agreed to speak first but said he would not get into a debate with a crowd that he perceived as possibly having a lynch-mob mentality.[5] He began by saying, "Beauty is in the eye of the beholder. While many people find wind turbines graceful and attractive, others disagree."[6] He addressed issues that Mountain Communities had brought up publicly in the local newspapers and reiterated that the twenty permanent jobs would not be minimum wage but would pay an average of thirty-five thousand dollars a year. The land around the towers would continue to be available for hunting, fishing, and recreation. The closest any home would be to a tower would be more than a mile away, and most homes would be at least four miles away. Wind farms didn't hurt tourism; they attracted more tourists. At a mile away, it would be unlikely for residents to hear turbine noise. And no endangered species had been seen in the area, so that was not an issue.

The meeting atmosphere was "tense."[7] Some people wanted to be reasonable; others didn't want to hear a word Groberg said. After he spoke, he headed out of the building. As he had said, he was not going to debate or take questions. Three men followed him out the door. Some of the Mountain Communities people worried there might be trouble. One went outside to make sure nothing happened and found out the men were part of the Invenergy group. The feeling was that, perhaps, Dave Groberg had never faced a "mildly hostile crowd" and wasn't used to one. It was thought that his experience was talking to people in West Texas, where windmills didn't much matter. After some discussion outside the community center, he agreed to return inside and take questions.

Meanwhile, John Stroud, representing Mountain Communities, gave his presentation. His title slide stated, "Wind Energy, Clean Energy or Corporate Boondoggle?" He said that his organization wanted "full disclosure of the facts" and a "full exploration of the unaddressed problems with wind energy." He talked again about subsides mentioning the accelerated depreciation for tax purposes that wind developments received and how lucrative wind plants were for their owners. Invenergy contended it would pay $400,000 in property taxes, but the county assessor said it would be $292,000. With regard to hunting and recreation, how could those "huge machines" not affect these activities? Tourists come to Greenbrier County for the scenic, natural beauty, not to view an industrial complex. Noise from "these things," he asserted, is driving people from their homes in Pennsylvania. Endangered species do reside in the area, including the northern flying squirrel, the Indiana bat, and the Virginia big-eared bat. The viewshed will be impacted with the turbines being seen from the counties major highways. There was a fire risk from lightning striking the turbines. A slide showing a turbine burning brought gasps from the audience.[8]

John Stroud made some other points. In hilly country (such as Greenbrier County), many existing roads were inadequate for getting the tower sections

Battle Begins

and the huge rotor blades into the area. He expressed concern about erosion, the disruption of water flow, destruction of wild habitat, and, once the towers were installed, the permanent impact to the area. "Wind power stations are not parks," he said. "They are industrial and commercial installations. They do not belong in wilderness areas. It makes no sense to tackle one environmental problem by creating another." He pointed out that "because of the intermittency and variability of the wind [impacting the flow of electricity from wind turbines] conventional power plants must be kept running at full capacity to meet the actual demand for electricity." They can't be turned off and on as wind energy fluctuates because they generate more pollution when started up.

Stroud concluded his presentation: "Beauty is in the eye of the beholder, and for once we [Mountain Communities and Invenergy] agree," as he showed a slide of a bucolic farm with a ridge in the background followed by a slide of a row of turbines lined up on top of a ridge.

ENTREPRENEURSHIP

One person opposing the Beech Ridge project at an early Mountain Communities meeting said, "There is a guy in Chicago who's not going to make $80 million if this [wind farm] doesn't go through."[9] That guy was Michael Polsky, an entrepreneur who founded Invenergy in 2001, and he had already made millions. The company, as an independent energy provider, operated in the competitive energy market using wind turbines as well as conventional power plants burning natural gas, coal, oil, and waste fuel. Invenergy was third in a series of companies that Michael Polsky formed; he sold his first company to a partner, and the second he sold after ten years to Calpine Corporation, also an independent energy company, in a $450 million deal. He stayed with Calpine briefly, then resigned and started Invenergy. His timing was excellent; Calpine declared bankruptcy in 2005.

Polsky, a trim, gray-haired man, immigrated to the United States from Russia in 1976 with five hundred dollars in his pocket. He found an engineering job with Fluor/Daniel, a construction company, and while there he earned a master of business administration degree from the University of Chicago. The *Chicago Sun-Times* portrayed him as arriving in this country knowing only a few words of English and as having become "one of Chicago's most successful entrepreneurs." A highly publicized divorce from his Russian wife in 2007 indicated a net worth of about $350 million, of which half was to go to his wife. His Invenergy assets were valued at $75 million.[10]

Invenergy had wind projects in twenty states and in Canada and Europe. To support its development, the company had ordered over five hundred wind turbines from General Electric to be delivered in 2006 and 2007. Its website declared that "Invenergy Wind takes a proactive approach to building strong relationships with various stakeholders including landowners, host communities, and power purchase clients."[11] Invenergy, as were other wind energy developers, was an exempt wholesale generator, or EWG, a classification created by Congress as part of the 1992 Energy Policy Act, which promoted competition by deregulating the electrical utility industry. A wholesale generator like Invenergy was exempt from certain financial and legal restrictions stipulated in the Public Utilities Holding Company Act of 1935. An EWG could sell to publicly owned municipal utilities as well as to whomever and at whatever rate it chose, without government regulation. Unlike public utilities, its rates were not regulated by the states.

Operating out of Invenergy's Washington, D.C., office, Dave Groberg led the effort to gain approval for the Beech Ridge project. In West Virginia, as the face and voice of his company, he made presentations promoting the endeavor and responded to concerns and criticisms of the project. He had always liked the outdoors, and he wanted to take a business approach to environmental issues. He received an MBA from the University of Texas and went to work for a wind energy developer in Austin. In 2004 he joined Invenergy. For him, the West Virginia project was a new experience. He had never dealt with the kind of opposition he faced there or any type of public controversy. Sometime later, Groberg would describe that meeting in Williamsburg as "totally a setup." He had no idea there would be a presentation, and he didn't know the press would be there. He had approached the whole thing, he noted, "with a certain naiveté." He said that Mountain Communities wanted to stop the project and in a way he had helped them to do it. "It was not a lot of fun," he added.[12]

John Stroud remembered the meeting a bit differently. He said that Groberg had suggested in advance that they exchange presentations. He received Groberg's presentation, he said, but never sent his. Debbie Sizemore thought Groberg was not yet forty years old. She described him as a man with a job to do who was determined to get it done. "That's the fairest thing I can say," she said.

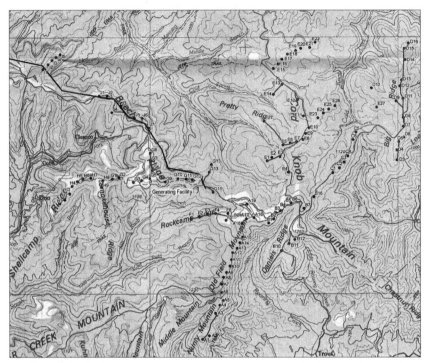

Invenergy planned to place 124 turbines along twenty-three miles of ridgeline. (Courtesy of John Stroud.)

JUDGE AND JURY

Looming over the developing battle between Invenergy and Mountain Communities as judge and jury stood the Public Service Commission of West Virginia. The PSC held the full authority to approve or disapprove the Beech Ridge project. It regulated companies providing electricity, natural gas, water, telecommunications, and sewer services in the state. Decisions were the prerogative of its three commissioners. Appointed by the governor, they served six-year staggered terms. No more than two could be from the same political party, and one had to be a lawyer.

Beech Ridge Energy would be the first wind developer to prepare a petition based on new rules established by the West Virginia Legislature that had become effective in July 2005. These "Rules Governing Siting Certificates for Exempt Wholesale Generators" applied to anyone constructing and operating an electrical generating facility that "sold electricity at retail outside the state or at wholesale in accordance with federal law preempting state law."[13]

On November 1, 2005, Invenergy, through their Beech Ridge Energy, LLC, subsidiary, formally applied to construct and operate a wind farm in Greenbrier County.[14] The company now proposed 124 turbines instead of the 133 they had previously talked about in meetings and presentations. Dave Groberg said that "after talking to some folks," Beech Ridge reduced the number of turbines and established a setback for the turbines, assuring that they would stay a mile away from any homes.[15]

The application stated that the wholesale market for the electricity generated by Beech Ridge would be for the mid-Atlantic region. To supply this electricity, Beech Ridge would have an interconnection and operating agreement with PJM Interconnection, a regional transmission agency that controlled the power grid for that region. To connect to the grid, Beech Ridge planned to run a fourteen-mile transmission line from the wind farm to an existing substation in adjacent Nicholas County. The 1992 Energy Policy Act, allowing competition in the public utility market, required utilities to allow other companies, such as Beech Ridge, to use their transmission systems.

Beech Ridge would use wind turbine generators made by General Electric, each with a rated capacity of 1.5 megawatts. The application estimated that based on the wind conditions on the site the project would generate electricity 85 percent of the time and, depending on varying wind speed, would operate in a year at 34 percent of capacity. Under these conditions, the facility would produce enough electricity to supply forty-five thousand homes, using on average 1,000 kilowatt-hours per home.

The project schedule called for installation of the turbines by December 30, 2006, "in order to attain project economies." To meet this schedule, the company requested that the Public Service Commission of West Virginia provide "expedited consideration" for the application by July 2006.

COUNTERATTACKS

Two weeks after Beech Ridge submitted its application to the PSC, Michael Woelfel, who lived in the Williamsburg area, filed a "Petition to Intervene" with the commission.[16] Woelfel stated that he owned thirty-eight acres on

Cold Knob Mountain and his property was "adjacent to, or within the parameters of, the one mile radius generating facility map filed by the applicant." He had "a clear view of the scenic overlooks and projected views, which would be adversely impacted by the proposed project." He further stated that the "industrial park" would impact the site of a "culturally significant event." He alleged that the Williamsburg area was the scene of an 1862 Civil War battle known as the Sinking Creek Raid. Also, he noted, the project "lies within the view ship of the Monongahela National Forest." Woelfel, concerned about noise, earth vibrations, water runoff, and property value, concluded that the "subject project, if approved, would constitute a private nuisance with respect to the personal and property rights of Petitioner."

The letter-writing campaign to the PSC began, accompanied by what might be called a media blitz by means of letters to the editors of the two Greenbrier County weekly newspapers. John Walkup III of Williamsburg wrote that the wind facilities would not exist if it were not for tax breaks, that they were "creating havoc" in communities, and that they generated very little useful energy. "They haven't replaced the first fossil fuel or nuclear plant nor reduced fossil fuel emissions," he said. The project, if approved, would not be "unlike the biblical Esau selling his birthright for a pot of soup." He believed that Invenergy would receive millions in tax benefits.[17]

Some letters raised questions about the MeadWestvaco property where the windmills were to be situated. Eddie Fletcher wrote about "the highest and wildest mountains in Greenbrier County and the enormous and horrific clear cuts practiced by MeadWestvaco, an absentee land owner. . . . Exploitation by out of state industry has plagued the Appalachian people, our people and the natural beauty for decades. . . . Their unsustainable management has left an 80,000 acre tract of land north of Williamsburg in despair."[18]

A Lewisburg resident, Glenn McKinney, had a different point of view: "A windmill high atop Cold Knob would be no more unsightly or intrusive than a radio tower, a ski resort, or a scar from strip mining. There can be no more unsightly eyesores than the thousands of telephone poles in towns, on highways and blighting the countryside." He compared the wind project "to a co-generation coal-fired plant belching tons of earth-warming, health damaging contaminants into the atmosphere, earth, and streams daily." He was referring to the co-generation plant planned for nearby Rainelle. "We must get away from these eco-damaging power sources sometime, and we'll never be able to succeed with this not in my backyard attitude," he said.[19]

Mountain Communities claimed that as of December 2005, through its letter-writing campaign, seventeen hundred letters had been filed with the PSC and only nine of those supported the project.

The Greenbrier County Convention and Visitors Bureau met at the Greenbrier Hotel to discuss the proposed wind farm. The organization was a force in the county, possessing a million-dollar budget largely supported by the bed tax generated by the hotel. The organization not only promoted the area but also subsidized a daily nonstop Delta Airlines flight from Atlanta to Lewisburg during the summer and fall seasons. Bureau director Kim Cooper summarized the organization's position: "The proposed wind towers would detract from our product, Greenbrier County's beautiful scenery." Lewisburg mayor and board member John Manchester challenged the board to determine if the project "helps or hurts" the bureau's mission to grow tourism in the neighboring Greenbrier Valley. Mayor Manchester said he had spent a lot of time in the Cold Knob area, and it was an area he encouraged visitors to see. "That's a very pristine area. I don't buy the argument that the wind turbines will enhance our ability to market the area."[20]

Gary Smith, present at the visitor bureau's meeting, said, "I am sadly amazed at the negative reception given to this project." He objected to the mayor of Lewisburg and the bureau speaking for Greenbrier County, especially the western end. Smith didn't think much of the MeadWestvaco land as a tourist attraction. He had spent "many years hunting, fishing, ramp-digging, ginseng-digging and other outings in the Cold Knob area" and wanted to know where the "pristine area" was. "What you'll find," he wrote, "is hundreds of thousands of acres of clear-cut acres and old/recent strip-mining. . . . It looks like bombed out wasteland."[21]

The Greenbrier Hotel and its Sporting Club residential development spoke against the Beech Ridge project. The hotel's president, Ted Kleisner, wrote to the PSC: "We believe we have a legitimate interest in any project that might impact our property, our business, or the quality of life in Greenbrier County that is so much a part of the experience of visiting The Greenbrier. Kleisner felt that the project has not been fully studied, and the developer failed to disclose the full visual impact of the project on the county."[22]

Greenbrier County state legislators Tom Campbell and Ray Canterbury entered the battle, placing before the legislature a bill repealing state tax credits that favored wind farm developers. The existing law, written not specifically for wind facilities but for any project that reduced pollution, taxed those projects at 5 percent of fair market value versus 60 percent for other industrial property.

Delegate Tom Campbell said, "If the turbines are going to be placed in Greenbrier County, then the tax monies received should offset any negative economic consequences of the project. Currently this type of property is being taxed at salvage value. You have to ask if this project makes economic sense to Greenbrier County." He looked at wind maps and felt that West Virginia was a marginal state for wind production compared to other areas of the United States.[23] (The American Wind Energy Association ranked West Virginia thirty-second among the states in wind energy potential.)

The two legislators' bills never got out of committee for a vote. It was felt by sources in Greenbrier County that heavy lobbying by MeadWestvaco and the Building Trades Council had exerted pressure to assure that the bills would not be considered. If the law had passed, Beech Ridge, as Mountain Communities had already pointed out, would pay $3 million in taxes to Greenbrier County instead of the estimated $400,000. Invenergy's Dave Groberg, in the mode of responding to every criticism of Beech Ridge, noted that states generally provide some kind of property tax relief for wind projects because of the high capital cost compared to other forms of power generation. And with wind being free, most of the costs go into the construction of the projects.

While the Greenbrier County Commission had no say in the approval of the Beech Ridge project, two of its three commissioners spoke for it. Commissioner Brad Tuckwiller, who lived in Lewisburg and belonged to one of the county's pioneer families, said, "Personally I find them [wind turbines] aesthetically appealing." He had toured Invenergy's seventeen-turbine site near Oak Ridge, Tennessee, while visiting his son at college. Commission president Betty Crookshanks was "excited about having the turbines near her Rupert community. . . . They would help the western end of the county," a less prosperous area that supported the project because of the promise of jobs. She said that the county had no ordinance that would prevent turbines from being constructed.[24]

EXTENSIVE INTEREST

As the opposition to the wind farm grew, Beech Ridge asked the Public Service Commission of West Virginia to hold a hearing in Greenbrier County and to make available the application documents they had filed with the

Battle Begins

commission. "The project opponents have conveyed misleading and inaccurate information on this project, and we would like to correct the record," Dave Groberg said.[25] Mountain Communities stated that it was they who had asked for the hearing, not Beech Ridge.

The Beech Ridge hearing took place in April 2006 in Lewisburg at the state fairgrounds. The three PSC commissioners sat on a stage at the head of a large room. The public sat in folding chairs or leaned against the back wall. Chairman John McKinney opened the meeting by saying that the hearing was being held because the case had "generated extensive public interest." This would be the only day to receive public comment before the board, he noted. Evidence in the case would be received in hearings in Charleston a month later.

Four hundred people attended the session, which went into the evening. Many of the people who spoke were from the rural areas that would be most affected by the view. A mix of beards, ponytails, dungarees, and a few suits characterized the crowd. Many people wore badges showing a windmill with a slash across it. A bus brought project supporters, presumably union members, from Charleston.

During the afternoon session, Groberg sat on the dais and was given ten minutes to present a synopsis of the Invenergy position before the public comment. In the evening session, Mountain Communities made a ten-minute presentation. Members of the public were allowed three minutes each to speak, and 150 people took the opportunity to do so. They were well spoken, mostly soft in their speech, and in some cases elegant in what they said.[26]

"You feel as close to God as you'll ever feel. Never in my wildest imagination did I think we would be overwhelmed by an industrial complex," said Dan Zahorenko, a ten-year county resident. "I don't want to see eight or more of those turbines filtering the sunshine. This is West Virginia's most beautiful land. With forty-story turbines, it's just a sad, a horrible thing," said Grif Callahan, who had lived for forty years near Cold Knob.

"Atrocious" was a word used by some. "Come hike in the woods, hear the birds sing," one speaker said, inviting the PSC commissioners to come see the site. "I may look ignorant, but I am not," one opponent noted as he began his comments.

"The wind farm is Einstein idiocy. . . . West Virginia was the laughing stock of the nation for its large satellite dishes. Put the turbines up in Chicago [Invenergy's headquarters], the 'windy city,'" said an eighth-generation resident of Cornstalk. "Farmland green and as pretty there can be," said another resident describing the Williamsburg Valley.

"Remember when West Virginians signed away their mineral rights?" said tourist consultant Judy Deegans, comparing those times to the present situation. She also noted that in her visits to tourist trade shows, no one asked where they could find wind turbines.

On the front page of the *Charleston Gazette* there was a photo of the hearing showing a woman with gray hair and a cane, sitting with a turned-down mouth. A framed picture of her two-story, log mountain home leaned up against her knees.

"The permit is seriously flawed. It pretends there is no impact [to the view]," said a man who lived half a mile from a proposed turbine. He said the viewshed map showed no impact to his property.

A Mountain Communities representative pointed out that the application to the Public Service Commission of West Virginia was indeed flawed. It left out historical sites, homes, and cemeteries on the site map as well as showing a viewshed that minimized the area from which the towers could be seen. Groberg admitted that the application was not "100 percent perfect" but said the mistakes were not intentional.

Some speakers spoke in favor of the Beech Ridge project, especially those from the western part of the county and the communities of Rupert and Rainelle. Gary Smith, from out that way, presented a petition signed by 217 people in favor of the project. He stated that the additional tax revenue and jobs would provide a "positive impact to our communities."

A few days after the Lewisburg hearing, Beech Ridge issued a press release announcing the results of a poll on wind energy sponsored by the West Virginia Manufacturing Association and wind industry representatives.[27] (The *Beckley Herald-Register* described the West Virginia Manufacturing Association as one of the state's most powerful lobbying organizations.) The poll indicated that 70 percent of the respondents favored wind farms in their county. Almost 80 percent said they were concerned about the need for more sources of power, and 90 percent thought that wind power was environmentally safe. A public relations firm from Charleston had contacted 501 registered voters and asked each one over twenty questions. Greenbrier County residents were included in the survey, and it was stated that their responses were largely consistent with the opinions of the statewide results, but no specific data was released. The report did present data on Tucker County, noting that 81 percent were in favor of the existing wind farm. When a copy of the report was obtained, it showed the Greenbrier data. The results indicated that in Greenbrier, in response to the question "Would you favor or oppose the construction of a wind farm in your county?" 64 percent were "strongly" or "somewhat" in favor.

The next step in the Beech Ridge approval process was the Public Service Commission of West Virginia's hearing in the state capitol of Charleston. The contest would become more complex, with Invenergy and Mountain Communities for Responsible Energy battling back and forth and the state government sitting in the middle.

Chapter 5

Rough Road

AN EVIDENTIARY HEARING

Charleston, West Virginia, with a population of fifty thousand, lies in a narrow valley along the Kanawha River about one hundred miles northwest of Greenbrier County by way of I-64 and I-77. Coal barges move downriver to power plants on the Ohio River. Coal trains roll eastward to seaboard markets. A few high-rise office buildings and the gold-leaf dome of the capitol dominate the skyline. The Public Service Commission of West Virginia is located in a renovated office building in the downtown.

The evidentiary hearing for a siting certificate for Beech Ridge Energy began on May 10, 2006. The hearing room looked like a small movie theater with raised seating for perhaps seventy-five people. The attorneys, the staff, and the intervenors arranged themselves at rows of tables "like at the UN," as one observer said. The three commissioners sat on a raised platform in front.

A Lewisburg resident who attended the hearing noted the body language between the PSC staff and those in the Invenergy group.[1] He said that it was almost embarrassing. They sat close together, almost as if they were playing a game, patting each other on the back and elbowing each other. Invenergy had a group of lawyers and assistants, well prepared and bearing rolling files. They brought in expert witnesses, spending an estimated thirty thousand dollars a day during the hearing. The commissioners were impressive in their patience.

Beech Ridge presented witnesses on birds, water, bats, the viewshed, sound, the power grid interconnection, cultural resources, wetlands, tourism, property values, MeadWestvaco support, turbine locations, and wildlife issues. Mountain Communities countered on each issue, emphasizing the viewshed and landmark issues. They presented their own five-mile map, a video, viewshed photo presentations, and a dark skies study. Individual intervenors, some associated with Mountain Communities, talked about turbine location, noise, health as related to bats killing mosquitoes, property values, and the effect on historic sites. The Building Trades Council union submitted evidence on the project's economic impact and on the use of local labor.[2]

Dave Groberg made some concessions: He guaranteed that Beech Ridge would pay for the removal of the turbines when they were no longer operational. The company would guarantee in writing that they would pay a minimum of $400,000 in property taxes. If taxes dropped below that level, they would make up the difference with contributions to the school board or some other county entity. He affirmed that hunting, fishing, and ATV (all-terrain vehicle) routes would be available on the property. Post-construction bird and bat studies would be conducted. And he said the site was one of the best spots in the state for wind. He continued to press his points as he had for almost a year—the jobs benefits, the $400,000 tax benefits to the county, and the fact that the PSC had some of the most thorough guidelines in the country. In summary, Beech Ridge presented a position that the company had the "expertise to construct and operate the facility," the "eastern United States needed additional electricity," and "Greenbrier County and Nicholas County and the entire state would benefit from the project."

Mountain Communities responded by saying that Beech Ridge had "overstated the capacity and need for the project . . . had overstated the economic benefits of the project to the state and the community, and that it had failed to disclose economic determents to the community that would result from the construction of the project." The commission had ninety days to render a decision.

A month later, West Virginia governor Joe Manchin proposed to a special session of the legislature that a six-month moratorium be placed on all wind projects within twenty nautical miles of any airport that handled military aircraft. "I was shocked it was put on the agenda," said Dave Groberg. Invenergy CEO Michael Polsky and Dave Groberg had met quietly with the governor in Charleston three weeks earlier. Groberg said that the governor had "never mentioned a moratorium, the FAA, or national security. Not once." Groberg went on to say, "It's a clear backdoor attempt to use the FAA issue to force the Legislature to re-examine the entire siting guidelines, even though they are basically brand new."[3] Meanwhile, he said his company was moving "full speed ahead."

The legislature ignored the governor's request. Why had he proposed the moratorium? The suspicion was that he knew it would never pass, but

Rough Road

it would make him look good to certain elements of the voting public while "any promises he might have made to the wind people would still be intact." The radar impact situation never materialized into an issue in West Virginia.

Back in Greenbrier County at a lower level in the political hierarchy, Dave Groberg and a Mountain Communities group each pitched their views on the wind project to the Lewisburg City Council. Based on a ten-minute time limit for each side, council member Tabatha Light said she didn't have enough information to take a stand. Member Beverly White was "uncomfortable" with being put in the position of having to make a decision. "It's a county-wide issue. People I talk to in Lewisburg don't seem to care either way," she said. The rest of the council was quiet when asked by the mayor if they wanted to make a motion.[4]

Always on the offensive, Groberg presented a check for ten thousand dollars from Beech Ridge to the Midland Trail Scenic Highway Association for a mural in Rupert. Association executive director Alice Hypes said that the interpretive mural would draw visitors from the mid-Atlantic states to the 180-mile byway crossing West Virginia and help market the route as a destination. Beech Ridge pledged another forty thousand dollars for more murals for other towns in Greenbrier County along the highway. The pledge was contingent upon construction of the $300 million wind farm, according to the *Valley Ranger*, Greenbrier County's free weekly.[5]

As Beech Ridge and its opponents maneuvered and the PSC staff deliberated, the commissioners canceled a hearing for another wind farm in the state. U.S. Wind Force had proposed a fifty-turbine project to be installed on a six-mile ridgetop in Pendleton County. However, the company refused to allow an opposition group's hydrologist to go on the property. There was concern that the forty-foot holes that would be dug to establish a base for the turbines would cause water runoff down the steep slopes into residential areas or into nearby limestone caves.

Was the PSC's cancellation of this project hearing an indication of what they would decide for Beech Ridge? Evidently not. On August 28, 2006, the Public Service Commission issued an order granting Beech Ridge Energy permission to construct a wind farm in Greenbrier County. The commissioners declared that they had had to "balance Beech Ridge's interest to construct a new generating facility, the state and region's need for new plants, and the economic gain to the state and local economy . . . against . . . the social and environmental impacts of the facility on the local vicinity, region, and state."[6]

With the order came a series of conditions. Construction should have minimal impact on the area, must start within five years, and must be completed by the tenth year. One line of turbines could not be constructed unless all property owners agreed. A fund must be established to dismantle the turbines and towers (when the time came). Beech Ridge must comply with the Endangered Species Act and file all environmental permits with the PSC. A three-year study of the impact on bat and bird activity must be conducted in consultation with a technical advisory committee. If the project caused significant bat and bird kills, then turbine operations must be adjusted (so-called adaptive management) to minimize the kills. And Beech Ridge must pay at least four hundred thousand dollars a year in taxes to Greenbrier County. All of the terms of the order applied to any subsequent owner or operator.

While these conditions reflected points raised by Mountain Communities for Responsible Energy and other intervenors, they also represented the wind turbine experience in Tucker County. Bat and bird kills there had been what many considered significant, and the issue between the county and project-owner NextEra over tax payments was ongoing

Shortly after the Public Service Commission of West Virginia issued their order, Al Gore's documentary, *An Inconvenient Truth,* came to Lewisburg's old downtown movie theater, a place that possessed the aura of a time when people had no worries about the environment or climate change. In rural West Virginia, the community often didn't have access to these types of films. The film's message was that greenhouse gases were accumulating in the earth's atmosphere as a result of human activities, causing surface air temperatures and subsurface ocean temperatures to rise. According to Gore, the evidence was "irrefutable." The hottest ten years on record had occurred in the last fourteen years, and 2005 had been the hottest. One of the film's strongest images was a map indicating areas of the earth that would be covered with water because of the melting of the polar ice cap. On another graph, emissions tracked the rise in temperature as if the two were a perfect match. *An Inconvenient Truth* won an Oscar for the best documentary at the Hollywood film awards in March 2007. Later that year, Al Gore shared the Nobel Peace Prize with the United Nation's Intergovernmental Panel on Climate Change (IPCC).

RECONSIDERATION

With the Public Service Commission of West Virginia's decision, disappointment reined among Mountain Communities members. But the organization responded quickly. On September 18, 2006, their lawyer, Justin St. Clair, issued a petition for reconsideration of the commission's August 28 order.[7]

He stated that Beech Ridge had failed to comply with the commission's own siting rules.

Beech Ridge's map was not to the specified scale and did not properly address recreation facilities, historic areas, or places of religious significance as required. Dave Groberg had even admitted some shortcomings with the map. Beech Ridge had not identified plans to mitigate the project's impact on significant sites and therefore had not done a cultural resource study. Mountain Communities had prepared its own map to point out the "insufficiencies" of the Beech Ridge map. The commission staff said that Mountain Communities' more detailed map offered sufficient evidence for the commissioners to make a decision. Mountain Communities called this "an absurd convoluted interpretation" in which intervenors had to supply information because the applicant had not.

The petition for reconsideration contended that the PSC had "improperly weighed the interests of other states" versus the interests of the citizens of West Virginia. Groberg's testimony stated that the project would help fulfill demand for renewable energy in the mid-Atlantic area, where several states had renewable energy goals. Beech Ridge was fulfilling the needs of other states by placing a renewable energy facility in West Virginia, a state that had no renewable energy policy. Wind energy, the petition continued, provides a "minimal" amount of electricity compared to other electric generating facilities. During the summer months, wind energy is least productive at the time when demand is the highest. Thus Beech Ridge failed to adequately justify the need for the project when balanced against the "potential long term adverse impacts" to Greenbrier County. Further, the commission staff, in noting that the viewshed was "the most hotly contested issue," declared initially that the removal of sixty-six wind turbines would "eliminate most of the potential negative impacts of this project." Thus the commission order "totally disregarded" the staff's suggestions.

With regard to bats, the commission findings were based on "inaccurate facts," the petition stated. Beech Ridge's bat expert had predicted that sixty-seven hundred bats could be killed each year by the project and that bats were killed at low blade speeds. Yet Beech Ridge would start the blades turning at a wind speed of eight miles per hour. In actuality, according to a letter from the U.S. Fish and Wildlife, the cut in speeds in Tucker County and at a site in Pennsylvania were ten miles per hour and should be fourteen miles per hour if bat deaths were to be reduced.

Mountain Communities asserted that the commission had overestimated the support for the project. The commission staff did note that over two thousand protest letters had been filed and had acknowledged that "it appears there is more opposition than support." And the commission order "seized" upon the towns in western Greenbrier County that supported the project because of the promise of jobs, yet they "represent the minority" in the county. With the majority against the project, the petition declared, the commission does a "disservice to the public" in granting the certificate.

Four months later, in January 2007, the PSC denied Mountain Communities' petition to reconsider.

TOO FAR APART

For the first time in a year, with the motion to reconsider pending, the proposed Beech Ridge project was receiving little attention from the media. However, John Stroud was doing more than playing music at Mountain Communities' fundraisers. Quietly, he and Dave Groberg had been meeting about the number of turbines affecting the viewshed.[8] Groberg offered to reduce the number of turbines by twenty-five. Mountain Communities wanted a reduction of fifty-four. The Invenergy offer would have deleted the ones closest to Williamsburg, including the one towering over the Stroud farm, but would have left twelve to fifteen in the upper end of the valley that would have an impact on people in the Friars Hill–Spring Creek Valley area. Groberg admitted that he could put ten of the deleted ones on the back ridges, which shouldn't be objectionable, but he would not concede more than the twenty-five number. Stroud did not think it fair to accept a reduction for the Williamsburg area and not include Spring Creek Valley residents to the northeast. Even with the fifty-four reductions, the wind farm would still have seventy turbines, far more than a lot of other wind projects. After four meetings, the two parties remained too far apart and negotiations ended.

Mountain Communities co-chair Debbie Sizemore did not support the negotiations for turbine reductions. "Some of us are die hard. I don't want one of them," she said. "I will disassociate myself from MCRE [Mountain Communities] if the negotiation on a reduced number are agreed to, but they haven't come to that point." Her father's ashes are scattered on Cold Knob. "That's as sacred as any place is. They [the turbines] are offensive—that they could put those monstrosities near where people's remains are." Sizemore said that this involved the whole community. "If I have to see one [turbine], I might as well see 124." To her it was not about viewshed and not about those people in Lewisburg but "about destroying a culture, a whole sense of community."

Rough Road

Table 5.1. Total Carbon Dioxide Emissions by Selected Country, 1990–2011				
Country	Billion Metric Tons			Percentage Change, 1990–2011
	1990	2009	2011	
World	21.7	30.2	31.6	46
China	2.3	7.7	9.0	291
United States	5.0	5.4	5.5	10

Sources: Guardian.co.uk, U.S. Environmental Protection Agency, Associated Press, International Energy Agency.

She mentioned Williamsburg's "Bear Dinner" fundraiser, which was in its forty-fourth year, as a part of the culture that would disappear. "There will be limited access [with the turbines]," she said. "It will stop hunting. They don't get it."[9]

With the negotiations between John Stroud and Dave Groberg having failed, and the petition for reconsideration denied, Mountain Communities directed an appeal to the West Virginia Supreme Court. It claimed that the PSC had not complied with its own rules and had not obtained adequate information about the project's impacts. In addition, Mountain Communities claimed they had been unable to cross-examine the reports of the state's historical commission and other agencies during the previous summer's hearings. The court granted a hearing, scheduling it nine months later, for January 9, 2008. The date fit Mountain Communities' strategy nicely—continue to delay construction as long as possible.

A UNITED NATIONS REPORT AND SOME SKEPTICISM

There may have been a lull in West Virginia while both parties waited for a state supreme court appeal, but on the national and international front, global warming became a media favorite. Discussions on its validity began in earnest.

The thunderbolt struck in January 2007. The UN's Intergovernmental Panel on Climate Change, in a report by scientists representing 113 nations, stated that it was 90 percent likely that global warming was caused by man. The atmosphere and the ocean had warmed over the last fifty years, and it didn't seem likely that all of it was a result of natural causes. A scientist's so-called hockey-stick graph showed a thousand years of steady global surface temperatures then, in recent times, a strong rise. These changes were already affecting environmental systems on every continent. The concentration of carbon dioxide had surpassed 350 parts per million and would reach 450 parts per million by 2100, which would increase the world's average surface temperature by 2 degrees Celsius (3.5 degrees Fahrenheit) and cause drastic climate changes. This was an increase these scientists believed would become critical, causing global climate patterns affecting rainfall, snow and ice cover, and sea level. In the millions of years of the earth's history, the concentration of carbon dioxide had never been so high. Prior to the industrial age some two hundred years ago, it was said that the level had been 250 parts per million, and the increase since then, they believed, was attributable to man and greenhouse gas emissions. The data was ominous, but in some scientific circles it was highly questionable, particularly in regard to what the concentration of carbon dioxide has been over the thousands and millions of years of the earth's history. However, at a Scripps National Ocean and Atmospheric Agency (NOAA) research station in Hawaii, the measurements of carbon dioxide in the atmosphere increased from under 320 parts per million in 1960 to over 380 parts per million by 2010. Again, some scientists said this was the greatest increase in history; others disagreed. This station is one site. Is it representative of a world condition?

Of the thirty-five billion metric tons of carbon dioxide emitted annually worldwide by the burning of fossil fuels, the United States emitted over five billion of those tons and at one point had the ominous distinction of being the world leader in emissions. But in 2007, China, with its growing economy, took the lead, quadrupling its output since 2000 as the world total increased 60 percent (see Table 5.1). Meanwhile, emissions peaked in the United States at six billion tons in 2007, declined as a result of the 2009 great recession, but rebounded with the economy in the two subsequent years. Then the effects of newly discovered, low-cost natural gas, which replaced higher-emission coal, took place, causing another reduction in emissions. (Coal dropped from 50 percent of total generation in 2005 to 34 percent in early 2012.) Ironically, the International Energy Agency stated that United States had cut emissions over the last six years more than any of the European countries, not because of government policies, but because of market forces. The Associated Press estimated that the United States' emissions would be

A coal-fueled power plant spews its emissions. (Photograph by the author.)

down to 5.2 billion in 2012, the lowest level in twenty years.[10] Was this a long-term trend and was it sufficient to affect climate change? Probably not, due to the instability in pricing and variability of energy resources and the fact that natural gas did generate carbon emissions, though much less than coal. And further, reductions had to occur globally.

Carbon emissions from electrical generation were 40 percent of the United States' total emissions. The balance came primarily from transportation. Europe, Japan, and India are the other big emitters. Underdeveloped countries will contribute an increasing amount of emissions as they become more industrialized. By 2030 some estimate that carbon dioxide emissions will exceed forty billion metric tons per year. Based on this emission level, scientists, using computer models, project the increase in the earth's surface temperature.

Carbon dioxide emissions are calculated in somewhat of a backward way. Figures are an estimate based on the carbon dioxide content of fossil fuels such as coal, natural gas, gasoline, and oil. The amount of these fuels used is measured, and a computer calculation determines the tons of carbon dioxide emitted. There is no measurement at the coal-burning plant's smokestack or at the exhaust pipe of a car. The measurement, as someone said, is like weighing the food you eat and calculating from that your weight. The numbers derived in this fashion can be manipulated. This became a concern at the climate conference in Copenhagen in 2009. The United States wanted to be sure that China's emission numbers could be verified.

The UN may have issued a report backed by a significant portion of the worldwide scientific community, but there was plenty of skepticism. The president of The United States, George W. Bush, indicated that global warning was too much of an unknown to establish goals or specific limits on greenhouse gas emissions. Scientists spoke out. Richard Lindzen, a professor of meteorology at the Massachusetts Institute of Technology, stated that global warming caused by humans could not be distinguished from natural causes. Patrick Michaels, a professor of environmental sciences at the University of Virginia, said, "What can you do about it? A lot of politically possible solutions will do less than nothing." William Gray, a professor of atmospheric science at Colorado State, agreed: "Humans caused the earth to be slightly warmer, but less than the [UN] report says." He bet that "in five to ten years the globe will start to cool."[11]

By 2012, the National Oceanic and Atmospheric Administration's (NOAA) Earth Research Lab was reporting 400 parts per million of "heat-trapping gas" over the Arctic. This was a breakout from readings worldwide, which had been in the 300s over the past sixty years. It had been 880,000 years since the earth had experienced levels in the 400s, stated NOAA. The conservative Competitive Enterprise Institute claimed that while carbon dioxide levels had increased, temperatures had not risen since 1998. But 2000 to 2009 was the warmest decade on record, according to NOAA records.[12]

NASA's James Hansen, a proponent for climate change action since the 1980s, issued another warning in 2012: "Global warning isn't a prediction. It is happening."[13] His near-term forecast of extreme weather events, ice-pack melts, droughts, floods, and the consequential economic losses were dire. Were his near-term forecasts coming true, considering the summer of 2012, with its extensive forest fires in the West, drought in the Midwest, and rare "derecho" storm in the mid-Atlantic states? Add to this list Hurricane Sandy, the huge storm that struck the New Jersey coast and New York City in the fall of 2012.

In the midst of the climate change debate, the National Academy of Science, an organization of experts who serve without fee "to address critical national issues and to give advice to the federal government and the public," issued a report. It expressed concern about global warming but stated that the nation's experience with climate change was based on a historic record of "relatively stable climate." The current efforts had been hampered by lack of solid information, uncertainty about the future, and a lack of coordination between government and scientific resources.[14]

IT'S NOT OUR JOB

While the world's scientists, including the UN and its skeptics, debated on the international stage the effect of carbon emissions on climate change, back in West Virginia in 2009, the issues between Beech Ridge Energy and Mountain Communities were strictly local. The state supreme court made its ruling.[15] In the hearing, Justice Brent Benjamin had set the tone early, saying, "We are not here to supplant the commission's decision with one of our own. It's not our job to make policy, it's our job to look at the law." Mountain Communities lawyer Justin St. Clair said, "Beech Ridge presented absolutely a minimal amount of evidence regarding the impact on cultural and historical landmarks," stating that the PSC had "essentially" written this impact out of their siting certificate. After commission lawyer John Auville told the justices that there would be a compliance hearing to assure Beech Ridge met a long list of conditions before construction could start, Chief Justice Elliott "Spike" Maynard asked, "Why are we having this hearing today then?" He repeatedly asked what the petitioners wanted the court to do. Jeffery Eisenbeiss, a co-petitioner speaking on his own behalf without a lawyer, responded by asking the court to dismiss the commission's order because they relied on Beech Ridge's experts and not on independent analysis. Judge Albright asked whether the issue was that the commission did not have sufficient expertise on their staff and therefore needed to supplement it with other experts.

Five months later, in June, the court denied the appeal and said that Beech Ridge could proceed with the project. The vote was four to one.

Construction begins on the Beech Ridge wind farm in Greenbrier County, West Virginia. (Courtesy of Mountain Communities for Responsible Energy.)

COMPLIANCE

Meanwhile, Beech Ridge began preparing the preconstruction compliance package to meet the eighteen requirements established in the original commission decision. They also had found a buyer for their power. They signed a twenty-year lease with American Electric Power (AEP) and its Appalachian Power subsidiary, which serves West Virginia, Virginia, and Kentucky. Though construction had not started, pending approval of Beech Ridge's compliance package, the agreement indicated that the project was expected to be on line by March 31, 2010. This lease was a part of the utility's goal to have 1,700 megawatts of renewable energy capacity on line by 2011, representing about 4 percent of its total capacity.

At the October 16, 2008, compliance meeting, Dave Groberg presented a nine-hundred-page document to the commission, the intent being that the site would be under construction the following spring. Five turbines had been eliminated because they were within one mile of thirteen homeowners living near Beech Knob Mountain.

Mountain Communities showed up at the hearing and asserted that Beech Ridge had not fully complied with the September 2006 commission order. They presented three expert witnesses to dispute several compliance items: an unsatisfactory agreement between the West Virginia Historic Preservation Office and Beech Ridge and what they defined as flawed studies and plans covering turbine noise, decommissioning of the site, and safety.

Mountain Communities had already issued an intent-to-sue notice to Beech Ridge ten days before the compliance hearing. The issue: violation of the Endangered Species Act. The suit stated that the wind developer was planning to construct industrial wind turbines near caves in which the endangered Indiana bat was known to reside. They also filed a motion that the compliance proceedings be delayed while they pursued their suit. The commission denied the delay.

A COMPLETE SURPRISE

While the Public Service Commission of West Virginia was considering the compliance issues, Mountain Communities sprung a surprise. In December 2008, in a letter to the developer's legal counsel, they proposed a settlement agreement that "essentially enables Beech Ridge to construct sixty-three of the turbines . . . and in exchange MCRE [Mountain Communities] withdraws its opposition to the balance of the project, and agrees to refrain from further adverse action against Beech Ridge."[16] The fifty-three turbines to be canceled were primarily located on the easternmost and southernmost ridges, which were nearest to the bat caves as well as to the communities that had the most objections to the turbines. This left the balance of the turbines on the less controversial western ridges, where there were no serious objections. In fact, the people in that area wanted the turbines since there would be some jobs available to maintain them. The commission's staff had suggested a similar reduction as far back as April 2006, when the commission was studying Beech Ridge's application. In addition, John Stroud and Dave Groberg had had an informal discussion about such a reduction in the fall of that same year, and here it was two years later.

Beech Ridge responded the next day, saying that the settlement offer was a "complete surprise" and that Mountains Communities made "an offer which it absolutely knows will be rejected."[17] It also stated that reducing the project was "economically impossible." The following day Mountain Communities fired back with a letter to the PSC declaring that Invenergy had smaller projects, one in Colorado with forty turbines and one in Idaho with forty-three turbines, and thus it was reasonable to have a sixty-three-turbine project. Furthermore, Mountain Communities was "dismayed" that its attempt to reach common ground had been rebuffed in such a cynical manner.

CASE FINAL

On February 13, 2009, the PSC issued an order that all of Beech Ridge's preconstruction conditions had been met and authorized construction to begin. They refuted all of Mountain Communities' disputed items, noting, "The compliance proceedings were not an opportunity to challenge the information that the Commission evaluated in 2005-06."[18]

Mountain Communities wouldn't quit. Eleven days later they submitted a petition for reconsideration of the February 13 commission decision. On April 3 the commission issued a final order removing the case from the open docket. Mountain Communities latest petition had asked for no more than a different decision.

Dave Groberg, now a vice president of development for Invenergy, said, "This case . . . has lasted nearly four years, and has resulted in the Beech Ridge Wind Farm undergoing the most intense scrutiny ever given to a West Virginia wind project by state and federal regulators and the state's Supreme Court. It is our intention to move immediately forward with the construction of this project."[19] He said he hoped the company could have sixty-seven turbines operating by the first of the year. "We're not done fighting this," said John Stroud, Mountain Communities' co-chairman. "This isn't the end."

Construction for the first sixty-seven turbines began in April 2009. Scrub growth was bull-dozed and wide gravel roads were constructed along the ridgetops. The turbine assemblies themselves began to arrive, and by mid-October, white, half-completed towers of the E-string jutted skyward in clumps along the ridgeline on the northwest side of the project. On the gravel road from Cold Knob to Ravenswood, now widened, a lone completed turbine was in place and its blades moved slowly. Construction vehicles stood at its base. It looked northwestward along a ridgeline that fell off into a wide valley, covered not by heavy forest but by scrub from past timber cuts. On a far ridge, another string of turbines would be placed. As in West Virginia, wind energy was progressing across the country, particularly in the flat, windy plains of Texas and the Midwest.

Rough Road

Chapter 6

Wind in the States

TEXAS BOOMS, THE MIDWEST FOLLOWS

At the time the Beech Ridge project was announced in 2005, the United States had 9,000 megawatts of wind capacity. By the end of 2012 the figure had grown to 60,000 megawatts. This was equivalent to roughly twenty-five to thirty new nuclear or coal power plants. Growth had started slowly from the California boom days of the 1980s and had occurred despite down years, when production tax credits had not been extended by Congress. Of the top ten states in capacity, Texas was the leader, followed by six states in the Midwest and three in the West (see Table 6.1).[1]

Windmills producing electricity for farms got their start in the Midwest in the 1920s because of the region's wide plains and consistent winds. The geographical and weather conditions encouraged the rapid development of utility-scale wind farms beginning in 2000. The area, including the Great Plains and parts of the West, stretched from North Dakota south to Texas and from Wyoming in the west to Indiana in the east. As one developer noted, almost the entire region had good wind conditions, was flat and thus ideal for construction, and was relatively rural, making land control easy. The state governments promoted the development with subsidies and renewable energy goals and landowners enjoyed the income from turbines being placed on their farms. The one common problem was how to get wind energy from remote areas to urban centers where it could be used.

The largest producing states in the country were Texas and Iowa. Texas had more than double the capacity of any other state. There were so many wind turbines in one area of Texas that the mayor of Sweetwater, Greg Wortham, said, "You can drive 150 miles along interstate 20 and never be out of sight of a giant wind turbine."[2] The landowners with turbines loved it. Johnny Usery owned twenty-five hundred acres of flatland with some cattle

Table 6.1. Top Ten States in Wind Energy Total Installed Capacity by Selected Year				
Rank	State	2005 (MW)	2012 (MW)
1	Texas	1,995	12,212
2	California	2,150	5,549
3	Iowa	836	5,137
4	Illinois	107	3,568
5	Oregon	338	3,153
6	Oklahoma	-	3,134
7	Minnesota	744	2,986
8	Washington	390	2,808
9	Kansas	-	2,712
10	Colorado	229	2,301
Total	(all states)	9,149	60,007

Source: American Wind Energy Association, Fourth Quarter 2012 Market Report, January 20, 2013.

and a few gas wells on it. With wind energy coming, he decided to change with the times and allow twenty-eight huge wind turbines on his property. "It took some soul searching," he said.[3] Turbines required less maintenance than cows; plus, he no longer had to worry whether or not it rained. Of course, Usery's bank account was also probably better off. Testimony in an Abilene lawsuit involving wind farms indicated that ranchers were leasing their land for as much as twelve thousand dollars per turbine per year. That seems an exorbitant figure; the more typical payment was a third of that per turbine. In later leases, developers offered a percentage of the revenue based on power purchased, in essence, a royalty similar to coal and natural gas payments.

Wind energy development in Texas began in 1999 with the state's incorporation of the Renewable Portfolio Standard, which established a target of 10,000 megawatts of capacity by 2025. That would be achieved by 2010. Ironically, George W. Bush, who was perceived as showing only mild interest in renewable energy while president, as governor of Texas promoted and signed the legislation that started the state's wind energy boom. The legisla-

Wind in the States

tion included the Renewable Energy Credit (REC) program, in which a wind energy company earned a credit for each megawatt-hour produced that could then be sold in addition to the electricity. Certain utilities in the state were required to buy a number of wind energy credits each year through a program administered by the state's grid operator. Texas governor Rick Perry summed up the wind energy situation in his state: "Texas has shown you don't need federal mandates to improve the environment or foster the next generation of energy technology." He expressed concern that a national cap and trade bill would not only impose a large tax hike but also inject the federal government further into every Texas home, farm, and workplace.[4] The governor seemed to forget that without the federal production tax credit, there would be far fewer wind turbines in Texas.

Iowa had a history of supporting wind energy development going back to the 1978 oil crisis, when it granted tax breaks to farm and business entities for wind and solar generation. By 1983 it had the country's first renewable energy standard. Beginning in 2005, the state offered a one-cent production tax credit in addition to the federal production tax credit. The Iowa Utilities Board allowed its two major utilities to increase rates in advance to pay for wind development. In 2008 Iowa quietly moved ahead of California in state ranking for wind energy capacity, and the two states have been neck and neck in second place behind Texas ever since.

At the annual 2010 Iowa Wind Energy Association gathering, Harold Prior, president of the association, announced, "Iowa has emerged as a world wide leader in wind power." The state was first in the nation in its percentage of total electricity produced and in wind manufacturing jobs, and second to Texas in total wind energy produced.[5] The Iowa Utilities Board had optimistically estimated that the state led the nation in wind generation as a percentage of total power output at 17–20 percent and had reached the benchmark that Denmark had established several years earlier. The American Wind Energy Association reported that in 2011 Iowa had actually produced 19 percent of its electricity from wind. North Dakota followed at 15 percent, and Minnesota at 13 percent. (South Dakota, with less than 800 megawatts of capacity, had generated 22 percent.) Texas, despite its huge lead in wind capacity, was at 7 percent, but it had a much larger power generation base than these smaller states.

With the prior expansion of wind farms in Texas, consumption from wind energy had doubled quickly. Turbines were being shut down when the wind blew because the grid couldn't handle the flow, and excess wind power had to be dumped. Such was the case on a February day in 2010. With strong winds blowing across West Texas, wind turbines operated briefly at a capacity factor of 70 percent, providing 22 percent of the power consumed in the state. (This situation had occurred in Denmark and Germany when peak wind generation at times exceeded the grid's ability to handle it.)

Across the Midwest, other states had taken aggressive steps to encourage the wind industry. Minnesota passed a mandate requiring that the utilities in incremental steps reach 20 percent renewable usage by 2020. Excel Energy, the state's biggest utility, had more stringent requirements: 15 percent renewable energy in three years, climbing to 30 percent by 2020, with wind energy being at least 25 percent of that. Kansas imposed a standard of 20 percent renewable energy by 2020; Illinois, 25 percent by 2025; and Colorado, 30 percent by 2020.

Despite having less wind potential than its neighbors to the west, Illinois had a rapid growth in wind energy capacity. It had a large demand center—Chicago—plus access to two transmission systems. Wind turbines were clustered on farmland south and west of the city. The state not only was in the Midwest Independent System Operator (MISO) grid but at the western edge of the PJM Interconnection grid, which spread to the East Coast. Next door, Indiana shared the same advantages—big cities such as Indianapolis, with Cincinnati in close proximity, and both the MISO and PJM grids. Though not having a renewable energy standard and getting almost all of its energy from coal, Indiana's wind capacity exceeded 1,500 megawatts of capacity by 2012, most of which had come on line in the last three years. Indiana finally established a renewable energy standard in 2011: 10 percent by 2025.

BIG FARMS

The biggest wind farms in the country initially were in Texas. In 2009, the 782-megawatt Roscoe Wind Complex went on line. Turbines were spread across one hundred thousand acres covering four counties. E.ON Climate and Renewables, a subsidiary of the same E.ON that worried about wind energy expansion and its impact on the grid in Germany, owned the facility. The company had invested "more than" $1 billion on the project and had negotiated with three hundred landowners, mostly cotton farmers, to use the land. E.ON claimed this to be the world's largest wind farm and stated that its London Array at the mouth of the Thames would be the largest offshore wind farm. However, by 2012 California's Alta Wind Energy Center had become the world's first gigawatt facility, surpassing the rest with a capacity of 1,020 megawatts and heading towards 1,300 megawatts. The other largest wind farms were NextEra's 735-megawatt Horse Hollow Wind Energy Center and Catamount Energy Corporation's 505-megawatts operation, all located near Sweetwater, west of Abilene.

The honor for biggest wind farm in the world would have gone to oilman T. Boone Pickens's development in the Texas panhandle northeast of Amarillo, but conditions changed his mind. The Texas oil and gas tycoon had started a nationwide publicity campaign to promote wind energy so that natural gas could be diverted from power generation to powering vehicles. He committed $2 billion to buy seven hundred turbines from General Electric and was planning to spend another $10 billion on his 4,000-megawatt project, which was to be completed in 2011. He figured it would be cheaper for the federal government to provide wind energy tax incentives for another ten years than what they were doing for nuclear power. "Try everything," he said. "Do everything. Nuclear, biomass, coal, solar. You name it."[6] He felt the only thing that could really reduce the use of oil was natural gas. But a year later, in 2009, Pickens canceled the project. The United States and the world were in an economic recession. Natural gas prices had fallen dramatically, making wind energy less competitive, and financing was difficult, particularly to build the transmission line that the project would need.

At the opposite end of the huge Midwest–Great Plains wind corridor, North Dakota was preparing for an even larger wind farm than any in Texas. Hartland Wind Farm, LLC intended to build in stages a 2,000-megawatt, $7 billion wind farm in the northwestern part of the state covering 720 square miles. With the fronts that swept out of Canada over this area, three years of wind data indicated the potential for an unusually high capacity factor of 45 percent. In fact, due to the potential stress of the high winds on the General Electric 1.5-megawatt turbine, a larger turbine might be required. Four years into the project, the developer had secured leases from four hundred landowners. A transmission line carrying power eastward to Chicago was critical to the project but still in the planning process. This project, when completed in total, was projected to almost double North Dakota's 2011 wind energy capacity.[7]

WEST COAST, MEETING CALIFORNIA'S NEEDS

John and Iva Grabner owned a wheat farm close by the Columbia River Gorge in Washington's Klickitat County, where giant turbines were being installed. They could look out their back door and see "dozens" of turbines rising as high as a "30-story building" from their fields. Lying on the ground "like gargantuan erector sets" were units yet to be installed. This was all part of the 133-turbine Big Horn wind project by PPM Energy of Portland. Forty-six of the units were to be located on the Grabner's farm. "It looks beautiful. Very space age," Iva Grabner told the *Seattle Times*.[8] John Grabner was not quite as enthusiastic about the view, and he had been surprised by the "amount of blasting and land clearing involved" in the project. But "if you are a part of it," he said, "it looks good." The Grabners were to receive $160,000 a year in exchange for the turbines being on their property.

Wind energy had come to the Northwest. It, of course, had been present in California for years. But Washington and Oregon with time had become a source for wind power that was needed by California to meet its renewable energy goals, the most stringent in the United States. California had mandated under Governor Arnold Schwarzenegger that the state reduce 25 percent of its greenhouse gases by 2020. Associated with this was a deadline that by 2010, renewable resources must supply 20 percent of the state's electricity. Authorities were concerned that the goal would not be met. The California Public Utilities Commission took a more optimistic view. California had the renewable resources. The goal would be met. It was just a matter of when.

Wind and solar projects were stuck, waiting approvals. The lack of transmission lines to bring wind and solar power from remote desert areas to urban areas was a major problem. But there were other problems as well: questions about financing, the complexity of regulations, the potential for contract failure and delay, the aging of existing wind turbines, and a concern about solar technology. The death of migratory birds, a problem that went back to the 1980s, continued to raise concerns.

Though California ranked third among the states in total wind capacity, it had added only 600 megawatts in the previous four years prior to 2010, but in 2011 it added 921 megawatts and 815 megawatts in 2012. Yet its total of almost 5,100 megawatts had to increase to 20,000 megawatts by 2020, according to the California Wind Energy Association, if the greenhouse gas emission goal was to be met.[9] Wind wasn't making it, putting pressure on solar energy. That included distributed solar (a plan to install solar panels on one million rooftops) and large solar facilities that consumed large areas of land. California utilities preferred solar power because, typically, on hot days when demand was greatest, the sun was shining but there was no wind. California's fiscal crisis left questions, certainly about achieving its wind energy goals. Getting wind power from Washington and Oregon would help, but put together the two states had just about the same amount of capacity as California.

Ideal wind sites lay on the eastern plains of Oregon and Washington in the vicinity of the Columbia River's Grand Coulee Dam. Here the usual problems—integration with the grid, the viewshed, and transmission line availability—were minimal. Wind power and hydropower complemented each other in feeding continuous, controllable electricity to the Pacific Northwest grid. When the wind blows, generating wind power, water flowing through the dam is reduced on a minute-by-minute basis. When the wind dies, water flow is ramped up to increase the hydro output. In comparison, ramping up and shutting down a coal-fired plant is slow and an inefficient process to compensate for wind variations. The Bonneville Power Administration believed that in this situation, where hydropower generated 67 percent of the electricity for the region, 6,000 megawatts of supplemental wind power (8 percent of the area's total capacity) could be easily handled by the grid. (In Europe, excess Danish wind power and Norwegian hydropower already had proven to be a successful combination, at least for Norway.) The "not in my back yard" issue was minimal in an area of semiarid, lightly populated farmland where property owners like the Grabners received two thousand to four thousand dollars per year for each turbine placed on their farms. In addition, local tax revenues increased significantly.

In an area where the timber industry had declined, crop prices were down, and an aluminum smelter had closed, the wind farm provided a positive for Klickitat County and was supported by the county commissioners. They established areas zoned for wind power, which simplified the permitting process. But in nearby Kittitas County, the commissioners rejected a wind farm while asking for setbacks to reduce the impact on property owners. Here there was local control, unlike in West Virginia, where wind energy development was in the hands of the state.

In Oregon, half the wind energy generated was being transmitted to California to meet that state's needs for renewable energy. To meet Oregon's 25 percent renewable power standard by 2025, Washington's 15 percent by 2020, and California's 33 percent by 2020, the *Oregonian* estimated that 16,000 megawatts of generation (at least 48,000 megawatts of capacity) would be needed on the West Coast.[10] This would be twice the amount of the existing federal hydro system (thirty-one dams and one nuclear plant). The paper called this "a staggering amount of electricity."

In 2008 the Oregon Energy Facility Siting Council approved the 845-megawatt Shepherds Flat Wind Farm. It would be one of the largest in the world, covering thirty square miles with 338 GE turbines, each with a

2.5-megawatt capacity. Developer Caithness Energy and its investment partners, Google and two Japanese firms, would pay $65 million per year in local taxes in two counties and $2.7 million in royalties for ten years to about two dozen landowners (an average of $112,500 per year per landowner). Farmer Loren Heidleman, one of the royalty recipients, said, "This is a big, big deal, not only for the landowners, but for the counties and all the people that are out of work." He failed to mention that it was obviously an attractive investment for the developer and the investors. Research by the *Oregonian* found a government memo stating that 65 percent of the total $1.2 billion investment cost was subsidized by federal, state, and local subsidies.[11] The $30 million from the state of Oregon came out to be $857,000 per each of the thirty-five permanent jobs that would be created. This wind farm opened in 2012 but came under criticism for the $30 million in triple tax credits that it received. The power from Shepherds Flat will not be used in Oregon but has been purchased by Southern California Edison to help meet the neighboring state's renewable energy mandate.

In southern California's windy desert, Southern California Edison had announced at the end of 2006 that they would buy 1,500 megawatts of power annually for twenty-five years from a wind farm proposed by an Australian financial group. The project would be located in Tehachapi, Kern County, in a fifty-square-mile area where smaller projects had existed since the 1980s. This was the last area in California for significant wind energy growth, as Altamont Pass and San Gorgonio Pass had run out of space. This wind project would increase the state's installed capacity at the time by 65 percent, but the original developer had gone bankrupt the year before. Terra-Gen Power of New York stepped in, bought control for $325 million, and this became the Alta Wind Energy Center with plans to eventually develop up to 3,000 megawatts of capacity in the area in several phases. By 2010, construction of the 150-megawatt first phase using 1.5-megawatt GE turbines was under way. For the next four phases, Terra Gen arranged a complex $1.2 billion financial deal in which the company developed the project, sold it to investors, then leased it back to operate. Those phases would add 570 megawatts of capacity using 190 large Vestas turbines with a capacity of 3.0 megawatts each. Another 830 megawatts would be added by 2015. These three increments totaled what Southern California Edison had committed to buy. An analyst with HIS Emerging Energy Research, Matt Kaplan, said, "The first Alta phases are very real, but future phases might be a little less tangible. We've seen California utilities sign up a lot of power purchase agreements for not the most realistic projects."[12] By 2013 the wind farm had an installed capacity of 1,300 megawatts, a little less than half the capacity originally planned.

In the East, wind capacity had grown slowly, reaching a capacity no more than what a couple of nuclear power plants could provide. And yet with its large population and power demands, the East was the region that could absorb the most wind power. Bringing wind energy from the Midwest over long transmission lines, however, would be inefficient and costly.

The northeastern states stretching from West Virginia to Maine had an overall installed capacity of 4,500 megawatts by 2013, an amount, if compared to single states, that would put the region fourth behind Iowa. New York and Pennsylvania had capacities over 1,000 megawatts. To the south, only Tennessee had any wind capacity, which consisted of one small wind farm. Two other states, Florida and Massachusetts, were significant factors in wind energy, yet neither had a wind farm operating within their borders. Florida had NextEra Energy Resources, the largest wind developer in the country, a former ambitious green governor, and a proposed wind farm that didn't make it. Massachusetts had been struggling for years to gain approval for the nation's first offshore wind facility.

New York, the East Coast leader, had the advantage of the brisk winds off Lake Erie and Lake Ontario. Yet its population and turbines at times did not mix well. Though near the major eastern markets, New York lacked the wide-open, low-population-density spaces that gave Texas and the midwestern states the advantage of more and larger wind farms. As one organization concerned about wind farms pointed out, Texas had fewer than one person per square mile in rural areas, while New York had a rural population twenty-five times that.

The state's renewable energy standard was 30 percent by 2015, having been increased from 25 percent in 2009. Existing renewables already provided 21 percent of that number, including 19 percent from hydropower, but only 0.5 percent came from wind.[13]

New York's largest wind farm—at 321 megawatts, the largest in the East—was Maple Ridge, seventy-five miles northeast of Syracuse. Its site was on the Tug Hill Plateau, a relatively flat upland recreational area exposed to constant winds off Lake Ontario. The project was a joint venture between Iberdrola Renewable and Horizon Wind Energy.

Noble Environmental Power owned the most wind farms in the state. Three had come on line in 2009, and the company announced an "innovative structure " of long-term financing to replace their construction loans. GE had supplied all the turbines, and GE Financial Services, with a portfolio of $4 billion in wind and other renewable energy assets, provided $200 million for the projects. A syndicate of banks and financial institutions, most of which were foreign, provided another $440 million.[14]

Noble made other news in 2009. On the morning of March 6, local residents of Altoona, New York, near Plattsburgh, reported a loud explosion with noises lasting several minutes. One turbine collapsed; another possibly caught fire and was damaged but did not fall. A loss of power to the facility initiated the problem. A few months later, in December 2009, another turbine collapsed. This occurred at the twenty-turbine Fenner Wind Power Project southeast of Syracuse. The 328-foot tower snapped at its base. The turbines had been manufactured ten years earlier by Enron and later acquired by GE Wind Energy. The project owner shut down the site and worked to strengthen the turbine bases by adding tons of additional concrete.[15] The final three wind turbines went back into operation eight months later, but the collapsed turbine was not replaced.

FLORIDA'S DILEMMA

Far to the south, Florida was home to NextEra Energy. It had two significant subsidiaries: Florida Power and Light, the state's largest utility, and NextEra Energy Resources, a diversified, nationwide wholesale and retail energy provider. The wind operations of NextEra Energy Resources had ninety wind projects in seventeen states and Canada with a total capacity of 8,570 megawatts—18 percent of the country's total. (At one point, it had one-third the total, but the Spanish wind developer Iberdrola was expanding at a more rapid rate.) NextEra's wind projects are clustered in West Texas, across the upper Plains states, and in California. In the East, there are five sites in Pennsylvania and one in West Virginia, the Mountaineer Energy Center. Ironically, there are none in Florida. In one early annual report, NextEra illustrated its emphasis on wind power with a picture of a little girl holding a colorful pinwheel. Inside the report, in a full-page picture, CEO Lewis Hay III posed in front of a series of wind turbines spread across a plain. At that time the company had 4,000 megawatts of capacity, and over the next four years it doubled that amount. Government subsidies, including the production tax credit, allowed NextEra to maintain a significantly reduced corporate tax bill.

Meanwhile, in 2007, the state's new governor, Charlie Crist, set Florida on a course to combat climate change. He held a "Serve to Conserve" climate summit in Miami. Governor Crist set the tone, warning the audience, "Global warming is real and Florida must be a leader in breaking the nation's addiction to fossil fuels."[16] He had been shown a map of the United States

Wind in the States

with the states marked that were taking action. The South was blank. His state was the third largest power consumer in the country and one of the top twenty-five producers of greenhouse gases in the world. The state's lengthy coastline was susceptible to not just hurricanes but also climate change and rising sea levels. The governor was sold. He presented goals to the state legislature: reduce greenhouse gas emissions 40 percent by 2025 and 80 percent by 2050 and call on utilities to produce 20 percent of their power from renewable sources, particularly solar and wind.

The Florida legislature passed a broad green energy bill in 2008. It directed that gas be part ethanol, cover conservation, establish tax breaks, encourage a possible cap and trade system, and provide some benefits for nuclear power and for new transmission lines. However, the governor's 20 percent wind and solar renewable standard did not pass. A legislative advisory committee determined that the state did not have the necessary wind resources compared to other states and that solar power was too expensive. A spokesman for one utility stated that if the governor's emission goals were to be reached, "the state absolutely has to embrace nuclear power."[17]

The governor did not get all that he wished from the legislature, but he was grateful to NextEra. The company had canceled a coal plant next to the Everglades while proceeding with one of the largest solar energy plants in the world (in California). And they were about to launch a wind project in Florida.[18]

In the fall of 2007, Florida Power and Light announced it would investigate the possibility of installing nine wind turbines along an undeveloped stretch of beach in St. Lucie County south of Fort Pierce next to its nuclear power plant. The company acknowledged that wind in Florida was not consistently strong and reliable enough to produce a large amount of electricity but wanted to "explore ways to best use this resource." The utility would draw on the expertise of its sister company, NextEra Energy Resources.

During the bus tour that Florida Power and Light and the St. Lucie County Commission conducted to the proposed beachside site, project development manager Henrietta McBee had spoken to the crowd at the first stop. The group then reboarded the busses and headed down A1A past the nuclear power plant and its two containment towers to adjacent property, where the turbines would be placed.

Four of the St. Lucie turbines to be placed on county oceanfront recreation and conservation land were rejected by the county administration. Florida Power and Light would shift three turbines from county land to unused state land, and six would go on its nuclear plant land. "They didn't solve anything,

Wind in the States

they just shifted the problem," said Commissioner Doug Coward.[19] With the continuing controversy, Florida Power and Light downsized the project to six turbines, all placed on company property. The utility touted the 22 million kilowatt-hours of emission-free electricity that would be generated. The project, at 13.8 megawatts capacity, operating at a capacity factor of less than 20 percent, would supply on average power to only 2 percent of the county's homes. The cost would be $45 million, or $3.3 million per megawatt of capacity. This was an enormous amount when compared to the $300 million Beech Ridge project in West Virginia (where a transmission line had to be built and construction would occur in mountainous terrain), which would cost half that—$1.6 million per megawatt of capacity. Three years later, a county representative described the St. Lucie project as "stalled." Commissioner Coward would have rather seen the utility and the state focus energy policy on rooftop solar panels.

Why had Florida Power and Light proposed this miniscule wind farm with its low efficiency, its high cost—to be paid for by its ratepayers—and minimal contribution to saving greenhouse gas emissions? Was it politics? Why had Governor Charlie Crist at his climate change meeting proposed wind energy as a significant factor in reducing greenhouse gas emissions in Florida? Even the American Wind Energy Association ranked Florida forty-seventh in wind capacity among the states. Florida Power and Light, while perhaps attempting to support the governor's position, basically weakened it. After searching the state, the company could come up with nothing better than a six-turbine wind farm that had to be put on the utility's own property. One Florida utility manager, who wished not to be quoted, saw its value as nothing more than a photo-op for the governor. The state legislature seemed to better understand how little wind capability Florida had. It had dropped the governor's requirement that 20 percent of power capacity expansion had to come from wind and solar sources.

CAPE WIND, THE OFFSHORE HOPE

Once off the mountain ridges or removed from the breezes off the eastern Great Lakes, and with little wind potential in the Southeast, the best wind resources on the East Coast are off the coastline. The first proposed offshore wind farm, Cape Wind on Nantucket Sound, became the nation's most controversial and publicized wind project, the classic NIMBY project. It consisted of 130 turbines with 468 megawatts of capacity covering twenty-eight square miles four to five miles off Cape Cod.

Jim Gordon, a New England energy entrepreneur, first proposed the wind turbines in 2001. He did not anticipate the furious opposition that arose. The wealthy and the powerful in the area—historian David McCullough, TV commentator Walter Cronkite, Chairman Jack Welsh of GE, and Senator Teddy Kennedy—opposed windmills being placed in the pristine waters of "their" Nantucket Sound. Suddenly this body of water became "a national treasure." An alliance formed that over the next few years spent $10 million opposing the project, while developer Gordon spent $30 million promoting it.[20]

Politically the situation was complicated. The federal government had jurisdiction, as they did with offshore oil wells. Responsibility shifted from the Corps of Engineers to the Department of Interior's Minerals Management Service, with Congress' Energy Act of 2005. This caused delays. In April 2010, nine years after it was first proposed, Interior Secretary Ken Salazar announced his approval of the project, stating, "This is the final decision of the United States of America." But it wasn't final; appeals were pending. U.S. Representative William Delahunt, a Democrat whose district included Cape Cod, declared that the project would "raise the region's power costs, disrupt an ocean sanctuary, and set back the wind-power industry, all to benefit a private developer."[21]

The cost of the project, about which no real information had been released, increased from the *New York Times*'s initial $1 billion to $2.5 billion by 2012. At the latest figure, the cost would be $5 million per megawatt. This reflected the general opinion that offshore facilities were twice the cost of those on land. Opponents estimated a cost in the billions of upgrades to the regional grid and to transmission lines. Cape Wind did reach an agreement to sell half of its power to National Grid, an international, investor-owned utility operating in the Northeast. This utility was subject to the recently established New England Regional Greenhouse Gas Initiative cap and trade system, which forced the purchase of a certain amount of renewable energy. It would pay Cape Wind 20.7 cents per kilowatt-hour, escalating 3.5 percent each year thereafter. National Grid said the impact to the average customer's monthly bill would be about a 2 percent increase. (According to the U.S. Energy Information Administration, the average retail cost of electricity in Massachusetts was about 16 cents per kilowatt-hour compared to 9.1 cents per kilowatt-hour for the country.)[22]

In mid-2012, Cape Wind received the supposedly final approval it needed: the FAA ruled that the array of turbines would not be a hazard to planes. Another utility agreed to buy an additional 30 percent of its power, putting the project in a position to arrange financing. However, there were still four

85

lawsuits outstanding against the project, and the project's nemesis since the beginning, the Alliance to Protect Nantucket Sound, pledged to appeal the FAA decision.[23] There was some optimism in Washington that construction might begin in 2013 after ten years of delay.

With shallow depths well off the coast, and with strong and steady winds, offshore wind farms provided the potential for significant growth for wind energy in the East. Several 300-megawatt facilities were in the planning stage: one off Rhode Island, three off New Jersey, and one off Delaware. And while they would be less likely to face the NIMBY problems that Cape Wind had, they did share the high-cost stigma. State regulators had caused a smaller Rhode Island offshore operation to not be approved by rejecting its price of 24.4 cents per kilowatt-hour as not commercially reasonable. The Obama administration through the Department of the Interior has encouraged offshore wind by establishing areas in federal waters for development as well as making the permitting process easier and shorter. That was all well and good, but the concern among developers was whether Congress would continue the tax credits and loan guarantees. However, if the East and Southeast were to reduce fossil-fuel emissions with wind energy, the main hope lay offshore. And while Europe had over fifty offshore wind projects along the coasts of ten countries, the United States had none.

Though the federal government had not established a renewable energy standard, many states had set standards that generally required wind to supply 20 percent of the state's electricity by a certain date. The politicians were good at setting goals but short on the details of how those goals were to be met. Wind is in competition with the other power-generating sources. Every source has its pluses and minuses.

86

Chapter 7

Alternative Sources

FOSSIL FUELS: CAN'T LIVE WITH THEM, CAN'T LIVE WITHOUT THEM

Wind as a source of energy has grown rapidly, but to continue that growth and to be a factor in suppressing climate change, it has to replace fossil-fuel emission sources, yet prove that it is the most advantageous of the non-emitting sources. Coal is the most obvious and the most challenging target as it emits the most carbon dioxide but remains the most entrenched energy source.

Roughly 600 coal plants in the United States annually emit one-third of the country's total carbon emissions, an amount equal to the emissions from the entire transportation sector. An additional 150 coal-fired plants had been in the planning stage according to the National Technical Energy Laboratory, but 30 of those had been canceled by the end of 2007. The cancellation number had increased to 83 by 2009.[1]

Two of those cancellations were Florida Power and Light plants. St. Lucie County, the same county that disapproved of wind turbines on its beaches, had earlier rejected a large coal-fired power plant. With that loss, FPL proposed another coal plant near the Everglades. This $5.7 billion, 1,960-megawatt operation was touted as having the most advanced coal-generating technology, thus producing more energy with less waste. In this case, the Glades County commission wholeheartedly approved the project. It would provide three hundred jobs plus property taxes that were almost twice the county's operating budget. But the Florida Public Service Commission killed it. It was concerned about the project's large construction cost, which would be passed on to consumers, and the risk of future fuel costs. Was the commission thinking, like Governor Crist, that Florida could get 20 percent of its power capacity from wind and solar energy? The utility was left in a quandary in a state and a nation with no energy policy. It responded by saying the coal cancellation decision would lead the company to build more natural gas plants, "putting all of its eggs in one basket." Half of Florida's power was already fueled by natural gas, 20 percent by oil, 20 percent by coal, and 10 percent by nuclear.

NASA's Jim Hansen told Congress in July 2008, "The only realistic way to sharply curtail CO_2 emissions is to phase out coal except where CO_2 is captured and sequestered."[2] And the phase-out should be done by 2030. Yet the UN's Intergovernmental Panel on Climate Change didn't see any significant use of that technology until 2050. Carbon capture had gained attention in Europe, but there were only a few small demonstration plants there. In the United States, President Bush had canceled a large demonstration plant in Illinois because the $950 million budget had been significantly exceeded. A pilot carbon capture and storage (CCS) project began operating in 2009 in Mason County, West Virginia. American Electric Power, at its Mountaineer power plant, was capturing a small stream of carbon dioxide from 20 megawatts of the plant's 1,300-megawatt capacity and pumping it 1.5 miles underground near the Ohio River. To advance the process to full commercialization by 2030 or even 2050, the high cost, the additional power needed for the process, and the land near a plant required for underground storage were considered drawbacks.[3] In addition, there was concern about the risk of storing large amounts of carbon dioxide underground.

In 2011 AEP canceled its carbon capture project. Blamed were Congress and the Obama administration for not having established legislation to regulate and reduce carbon emissions. Without this, the utility could not get state regularity approval to cover the utility's share of the cost. The federal government was funding half of the almost $700 million cost. The Public Service Commission of West Virginia agreed to allow AEP to pass on only a part of the remaining cost to its customers. Meanwhile, West Virginia politicians had fought any national emission reduction regulations.

Even with CCS far into the future, Kevin Book, an energy policy analyst, stated that there would be 16,000 megawatts of new coal plants coming on line in the United States in the next few years, but "they may well be the last plants."[4] In 2009 eight coal plants became operational, totaling 3,218 megawatts, the largest increase in capacity in eighteen years.

In West Virginia, where it was felt that the coal industry was under attack by the Obama administration, Secretary Stephen Chu of the Department of Energy, accompanied by West Virginia senator Jay Rockefeller, arrived for a

The Dominion Resources coal-fueled power plant at Mount Storm, West Virginia, provides power to the East Coast. (Photograph by the author.)

visit in September 2010. They had conducted a forum on coal carbon seques-tration at the University of Charleston. Senator Rockefeller led off. Coal was the 24/7 base load for power generation, yet it was at a crossroads. Climate change was real; carbon dioxide was not "healthy" (small patter of applause followed). It was not a myth; it cannot be willed away. We can't "put our head in the sand," he said. Renewables can't replace coal anytime soon and cap and trade was not the answer. It would be cheaper to develop carbon capture. "But there is uncertainty in the law; we need clarity about the tech-nology; we need 'rules of the road' in order to encourage investment," he stated. Secretary Chu, a Nobel Prize–winning scientist, expressed concern about climate change and optimism about what could be done with coal to reduce carbon dioxide emissions. Carbon capture development could reach a commercial state in eight to ten years, he believed. The world had lots of coal reserves and they would be used. "Let engineers work on it instead of lobby-ists [and] miracles happen," he said. The United States must lead, he added, or we will have to import the technology. But he closed with a projection: "In 20 years carbon dioxide will be obvious."[5] There were many who thought it was already obvious with every major weather event.

The most dramatic reduction in carbon dioxide emissions would come from replacing coal with renewable energy, Chu said. Wind energy costs (with its production tax credit) were approaching parity with coal, but coal provided base load electricity and had many years of low-cost reserves available. Could wind, with its intermittent production, ever become a base load supplier replacing coal? Certainly not on the short term, he concluded. Coal was here to stay, but for its long-term viability, the development of carbon capture was critical. In the meantime, the Obama administration through the Environmental Protection Agency was issuing rules to reduce noncarbon emissions from coal-fired power plants. But it was the sudden emergence of natural gas reserves that was having an impact on coal and, to a degree, wind energy.

THE FUEL OF CHOICE

While the University of Charleston forum was about coal, Senator Rockefeller warned that natural gas was becoming "the fuel of choice." He quoted the president of American Electric Power as saying that "the utility could switch from coal to natural gas tomorrow."

With its coal plants canceled, Florida Power and Light proceeded by rebuilding two aging natural gas plants and adding three natural gas units to its Martin County facility west of West Palm Beach. The utility submitted a ten-year plan to the Florida Public Service Commission. The utility needed to meet the demands of its customers by increasing capacity 30 percent by 2017. Of that total, almost a quarter would come from conservation (demand-side management) goals directed by the commission, including the use of more efficient air conditioners, the institution of load control programs, and by improving and adding home insulation.[6] The balance would come from added capacity, part of which included the combined cycle gas upgrades, new units, and nuclear upgrades, all in process. The rest of the capacity depended on retuning older units that were on inactive reserve status back to service. Two new nuclear units were planned but would not come on line within the ten-year time period. Though Florida Power and Light had three small solar operations, neither solar nor wind power was part of the plan. Mentioned as a factor regarding future planning was the possibility, unlikely so far, of

Alternative Sources

the state legislature establishing a renewable energy standard. Florida could well be on the path of 75 percent of its energy coming from natural gas and basically staying neutral in preventing climate change.

As in Florida, natural gas was becoming the most practical, near-term energy source. In two Internet energy reports, it was called the New King of Electric Power and the Fuel of Choice. "Coal has been removed in many places as an option," said Art Holland, a utilities consultant with Pace Global Energy Services.[7] The firm forecasted through 2025 only a moderate growth in coal-fired capacity but doubled its forecast for natural gas plants. New nuclear plants were questionable. Sun and wind power were growing but would provide only a "small part" of the needed capacity, plus they were intermittent. By "default" this country would have to "fall back" on using natural gas, Holland said. Natural gas, which accounted for 22 percent of electricity production in this country, had advantages over coal besides generating on average half the carbon dioxide per megawatt of capacity and having dramatically less mercury. The investment cost was less per megawatt of capacity; its plants could be built in smaller yet efficient increments, typically 500 megawatts versus 1,000 megawatts for coal; and these smaller plants could be distributed closer to markets than the larger centralized coal plants. This distribution provided more security, reliability, and flexibility. In 2008 natural gas reserves in the United States increased 35 percent from the 2006 estimate. The increase resulted from new drilling techniques and the discovery of new reserves. Using horizontal drilling and hydraulic fracturing (high-pressure water to fracture rocks and release trapped gas) allowed more gas to be tapped. The Marcellus Shale, underlying Appalachia, added significantly to reserves. However, initial drilling in the Marcellus and the fracturing of the shale had led to concerns about the effect on drinking-water supplies. Demonstrating the future importance of natural gas, oil giant Exxon bought XTO Energy, the second largest gas producer in the nation.

In the past natural gas was seen as a peaking source to handle quick increases in power demand over what coal and nuclear base load sources provided. It was also considered a backup for rapidly balancing and smoothing the load when wind fluctuations affected the grid. But by 2012, with its newfound reserves, stable low price (dropping from eight dollars to almost two dollars per unit), and reduced emissions compared to coal, it was having an impact on the growth of wind energy as well as the growth of other energy resources. The buzz was with wind and solar, but natural gas was becoming the practical choice. The United States received 52 percent of its electricity from coal in 2000. In ten years this decreased to 45 percent as

91

natural gas rose from 16 to 24 percent. (All renewables had risen from 9 to 10 percent.)[8] The Energy Information Agency reported that coal was at 34 percent in 2012. The cause: primarily natural gas.

NUCLEAR DILEMMA

Wind did not emit greenhouse gas, but neither did nuclear power, a fact that sometimes seemed overlooked. In 2010 the United States had 104 reactors located in sixty-four nuclear power plants, providing 20 percent of the nation's power. Yet the Nuclear Regulatory Commission had not approved the construction of a new plant since 1978. In 2010 the Obama administration announced $8.3 billion in loan guarantees for two planned nuclear reactors in Georgia. There were eight other reactors in the United States that were planned, but overall nuclear expansion was experiencing hurdles due to the always-present safety and waste concerns. But construction costs were also becoming a concern. Worldwide there were 443 operating reactors in 2010, providing 14 percent of the world's electricity. In France, 58 reactors provided more than 70 percent of that country's power. Per the World Nuclear Association, 220 reactors were under construction or planned. China, where the strong, central government easily made decisions, had only 13 operational plants, but 77 were being processed or built. However, because of age, many plants worldwide will have to be shut down by 2030 and will have to be replaced just to maintain the present capacity. In 2011 an earthquake tsunami struck Japan, causing the world's third major nuclear catastrophe. As a result, Germany decided to shut down all its nuclear plants by 2020; yet nuclear energy power at its peak had provided 20 percent of the country's power. The International Energy Agency announced that if nuclear energy expansion were cut in half, carbon dioxide emissions would increase by 30 percent through 2035.[9]

The safety of wind turbines tends not to be an issue unless a blade flies off a turbine, as happened in New York state. Nuclear safety has always been an issue. The world's worst nuclear disaster, caused by bad management and bad design, occurred at Chernobyl in Russia in 1986. During an equipment test to get more power, control rods were removed and coolant water was reduced. The chain reaction accelerated and the reactor exploded. The reactor had no containment building to hold back the radiation when the core meltdown occurred. In a study completed in 2005, the World Health Organization reported almost fifty deaths within months of the accident. The

Alternative Sources

study "estimated" that as many as four thousand people of the two hundred thousand who lived in the contaminated areas "could eventually die."[10] The 1978 accident at Pennsylvania's Three Mile Island resulted from a pump failure. The reactor core had a meltdown and gases containing a low level of radiation had to be vented. No deaths occurred and a state agency tracked thirty thousand people in the area for twenty years and found no sign of effects from radiation exposure. The 2011 meltdown of three reactors at Japan's Fukushima nuclear plant occurred as a result of the tsunami, which was triggered by an earthquake. Water flooded the plant, knocking out the backup power and the ability to cool the plant's reactors. The tsunami by itself killed twenty-four thousand people. One immediate death was attributed to the nuclear meltdowns, but with some radiation leakage, the long-term toll remains unknown. As the *Economist* reported, Fukushima was "nothing like as bad as Chernobyl" and "a bit" like three Three Mile Islands in a row.[11] Meanwhile, four hundred other nuclear reactors in the world have operated with no reported deaths from accidents. The nuclear power plants of today descended from technology developed for Adm. Hyman Rickover's U.S. Navy programs. Its reactors, powering aircraft carriers and submarines, never had a serious accident. Yet during an eight-year period from 2002 to 2010, almost two hundred men died in the United States from coal mining accidents. This figure includes the twelve deaths from the Sago mine in 2007 and the twenty-nine deaths from the Upper Big Branch mine in 2010, both in West Virginia.

Wind turbines don't have waste. Nuclear plants do, and, like safety, waste causes much consternation. At the St. Lucie plant and at other nuclear plants, the waste consists of highly radioactive, used fuel rods. They are initially placed in wet storage in an underground pool of water in the containment building. When that area is full, the rods go to dry storage, where they are placed in canisters and stored in concrete and steel modules on the plant property. The canister provides radiation shielding, air cooling, and protection from natural disasters like earthquakes or hurricanes. Dry storage provides protection for nuclear waste for at least fifty years and probably one hundred years, according to the Nuclear Regulatory Agency.[12] Nuclear waste has been stored at nuclear plants all over the country for forty years because the government has been unable to establish a national repository. Yucca Mountain was to be the location. But after years of investigations and tests, going back to 1982, and $10 billion in expenditures, it was canceled. The reason, in realty, was political, not technical. It was not as if the federal government had never built a repository. At Carlsbad, New Mexico, the

Waste Isolation Pilot Plant was in full operation. Nuclear military waste had been stored there since 1999. The state of Nevada, led by its senator, Harry Reid (also the Senate majority leader), did not want it. President Obama, during his election campaign, made a commitment to Senator Reid that he would cancel Yucca Mountain, and he did. The president established a blue ribbon committee to study the disposal of nuclear waste. The commission was not looking at new sites but at ways of reprocessing and reusing waste. Meanwhile, the Department of Energy continued to pay utilities millions to store waste on their own sites.

Reprocessing spent fuel reduced the space needed for nuclear waste storage. Other countries did it, but not the United States. The process produces plutonium that can be used in nuclear weapons. President Jimmy Carter, fearing proliferation, canceled reprocessing in the 1970s. Even without reprocessing for all these years, this country's one hundred nuclear reactors have generated very little waste. The National Energy Institute stated that if all the spent fuel rods produced in the United States were laid side by side and end to end they would cover only an area the size of a football field to a depth of fifteen feet.[13]

"Every source of electricity had its negatives and each raised questions," said John Holdren, a Harvard professor and authority on energy technology. "Nuclear power was too unforgiving of either human error or human malice. But were there enough reserves of oil and gas? Could the atmosphere tolerate the emissions from coal and tar sands? Were there enough good sites for wind and hydro? Would solar power always be too expensive?" He added a further question: "Can you solve the climate problem without nuclear energy?" He answered: "Yes . . . but it will be easier to solve it with nuclear energy."[14]

Despite Fukushima and the continuing concerns about safety and nuclear waste, people in St. Lucie County have lived comfortably with their nuclear plant for years. They walk the beach in front of the plant, picnic and surf nearby, and fish the swirls where cooling water for the plant is taken in and discharged back into the ocean. If ever a location illustrated the similarities and differences between two energy sources, the coexistence of the St. Lucie reactors and the planned wind turbines did. Their operating costs per kilowatt hour are lower than coal and natural gas. And in this case, being collocated, the wind turbines could piggyback on the nuclear plant's existing distribution system.

But otherwise the differences are great. The two containment buildings of the nuclear plant required minimal land. Because a wind farm of any size uses a vast area, the St. Lucie wind farm was limited to an impractical nine tur-

Alternative Sources

bines, at best, on the vacant beach due to development to the north and south of the plant. And there were objections even to that, because the existing land was used for recreation and conservation. The nuclear plant is not pretty, but it does not dominate the landscape to the degree the nine or even six wind turbines would have. Construction-wise there is a world of difference. A new reactor added to the site would require a huge investment (billions), a multiyear approval processes, and several years of construction, but it would provide a much greater amount of electricity. If the wind turbines had been approved, they could have been installed within a year. Even in large numbers, such as one hundred or more, the turbines can be constructed in a relatively short time. On a larger scale, with jobs an important issue in the aftermath of the 2009 recession, a nuclear plant requires many workers for both construction and operation, while a wind farm requires few. The two-reactor Georgia plant due for completion in 2017 would generate thirty-five hundred jobs during construction and eight hundred jobs when operating. An average-sized wind farm like Beech Ridge in West Virginia requires two hundred construction workers and fewer than twenty to operate it.

But in the end what really counts is the amount of emission-free energy produced. A wind farm's output is intermittent. A nuclear plant serves hundreds of thousands on a twenty-four-hour, constant basis. Wind variations make the grid difficult to manage. Nuclear energy, like coal, provides the power system's base load. Compare the Horse Hollow wind farm in West Texas, one of the largest in the world, with the St. Lucie Nuclear Power Plant. The wind farm's 421 turbines are spread across two counties covering forty-seven thousand acres. (Even in the wide-open spaces of Texas, some people have objected to the turbine noise as a nuisance but have been overruled in court.) The St. Lucie plant, using roughly 1 percent of the land required by the wind farm, has more than double the capacity (1,700 to 736 megawatts), has more than double the capacity factor (from 90 to 40 percent), and provides base load power to over 1.1 million homes on a consistent, twenty-four-hour basis. The wind farm provides power to an annual average of 220,000 homes on a variable basis (see Table 1.2). And the St. Lucie plant is located close to its customers, while Horse Hollow is far from its market, thus requiring a major investment in new, long-distance transmission lines.

As a result of the Fukushima disaster, both Germany and Japan have cut nuclear power and face the problem of how to compensate for the loss of power in light of their goals to reduce carbon emissions. Japan took all but two of their fifty nuclear reactors off line with no indication when some maybe restarted. This is a country that before the disaster had planned

A fishing boat cruises off the St. Lucie Nuclear Power Plant on Hutchinson Island, Florida. (Photograph by the author.)

to increase its nuclear power from 26 percent to 45 percent by 2030. The immediate result was a 17 percent increase in carbon emissions in the year after the disaster as more polluting crude oil fuels had to be used to supplement the lost power generation. One utility had a 40 percent increase in emissions. In an effort to stay on track to meet their 25 percent Kyoto goal emission reduction by 2020, they had to buy almost double the emission credits over the year prior to the disaster, increasing power company financial losses. While Germany was looking at offshore wind energy as one means to supplement nuclear power loss, Japan was considering an increase in renewables from 10 to 35 percent, plus new energy efficiency regulations. The government was also considering a range of plans, from no nuclear power to restoring as much as 25 percent of its nuclear power. As a Bloomberg.com article reported, most of the people were for the zero option.[15]

In taking into consideration both nuclear power's safety concerns and its practicality as a base load generator that does not emit carbon dioxide, Nils Diaz, the former chairman of the Nuclear Regulatory Commission and

Alternative Sources

a professor of nuclear engineering at the University of Florida, said that it would be impossible to regulate the industry to a "zero risk factor." The goal, he said, was to make sure there was "reasonable assurance of adequate protection." He also noted that the "world is going to go nuclear, because they do not have any other real alternatives."[16] And as a *Bloomberg Businessweek* headline stated, "Nuclear accidents like Japan's Fukushima are scary. So is a future without nuclear power."[17]

SOLAR: SUNNY OR CLOUDY

With Germany's plan to shut down nuclear plants, further growth in both wind and solar energy was viewed as a replacement. Increasing its installed solar capacity significantly in 2011 and 2012, the country reached a total capacity of 26,000 megawatts, almost matching its wind capacity and representing half of the world's solar power capacity. Most of this growth came from rooftop installations on homes, farms, small businesses, and large industrial concerns such as automobile manufacturer Opel, all while the growth in wind energy leveled off.

On a sunny Friday in May 2012, solar power met a third of Germany's electricity needs over the noon hour. Solar supporters were ecstatic. This was further encouragement to the politicians looking at renewables as a replacement for the shut-down nuclear plants.

Germany, despite being susceptible to cloudy days, had developed the world's largest market for solar cell systems. Subsidies equivalent to seventy-nine cents per kilowatt-hour were initially paid to homeowners and businesses to install panels on their roofs. Through this subsidy program Germany established a photovoltaic (PV) industry employing forty thousand people. As some German politicians declared, the program had become too successful for its own good. Power companies had to purchase the excess energy at a fixed price for twenty years. This mechanism, a feed-in tariff, provided a great incentive to homeowners, though utilities were allowed to pass the cost on to its customers, industry and homeowners. Household electricity rates in 2011 were thirty-six cents per kilowatt hour (with the value added tax, or VAT), a close second in Europe to Denmark with its heavy reliance on wind energy. France's rates, with its high percentage of nuclear power, were half those of Germany. And Germany's costly solar program provided 2 percent of its electricity generation in 2010 compared to 6 percent for wind. For the country in that year, wind's capacity factor was 17 percent, solar's

8 percent.[18] A year later, with its growth, solar power was producing 4 percent of Germany's electricity.

Solar energy, of course, does not produce electricity at night, and it barely produces during Germany's cloudy winters, thus leaving it with the low annual capacity factor. And like wind, it must be backed up with expensive conventional power. Meanwhile, homeowners and industrial operators with rooftop systems received $11 billion in payments in 2011. This plus other government subsidies were twice as high for solar as for other renewables. For the same amount of subsidy, wind produced five times more energy than solar. And Germany wanted to double its solar capacity to meet electricity needs as well as meet its emission reduction goals. Germany's *Spiegel* reported that an economic study showed that to reduce carbon dioxide emissions by one ton, it would cost six dollars to insulate an old building, twenty-five dollars to build a new gas-fired power plant, and six hundred dollars for photovoltaic arrays, considering subsidies and the still high though decreasing cost of solar panels.[19]

Following as world leaders in solar capacity, Italy, with a tremendous surge in 2011, expanded its capacity to 13,000 megawatts; Japan, Spain, the United States, and China were in the range of 5,000 to 3,000 megawatts. All had much better sun conditions than Germany but, like Germany, had strong solar incentives.

Spain, in particular, established itself as a strong solar energy competitor, as it had done in wind energy. However, its economic problems in 2009 caused the incentives to be reduced. Unlike Germany, Spain's emphasis was on solar thermal energy rather than solar cell energy. Solar thermal, or concentrating solar power (CSP), directs the sun's rays with mirrors onto pipes, heating a fluid that produce steam and drives a generator in the same manner as fossil fuels and nuclear energy. On the other hand, a solar cell, also known as a PV cell, converts the sunlight directly into electricity. Sunlight strikes a cell of semiconductor material that initiates a small flow of electricity. Cells are grouped in modules and mounted as panels for rooftop or utility-scale installations. Electric generation is dependent on the sun's intensity and the number of hours per day the sun shines.

Italy's 9,300-megawatt spurt in capacity in 2011 was triggered by a special government decree but was to be cut significantly in 2012 to reduce customers' power bills, which were supplementing the renewable surge. Italy, like Spain, was also facing economic problems. Solar cell investment in Spain was expected to be cut more than 90 percent from its 2010 peak.

In Asia, Japan's solar growth accelerated with reduced equipment costs and new government incentives. China and India, with minimum capacity in 2009, each had plans to expand to 20,000 megawatts by 2020. China became the world's leading low-cost supplier of photovoltaic cells, providing destabilizing competition for European and U.S. solar production. India in 2012 commissioned with much celebration the world's largest solar plant, the 600-megawatt, three-thousand-acre solar cell Gujarat Solar Park. The European Photovoltaic Industry Association stated in its market outlook for 2016 that Europe had dominated the market for years, but China and India, followed by other parts of the world, would have the fastest capacity growth in the future.[20]

In the United States, the Solar Energy Industries Association indicated that solar capacity overall grew from 500 megawatts in 2000 to 3,000 to 4,000 megawatts (depending on the source) by the end of 2011, yet at a pace over the period far slower in growth than wind. However, with the cost of solar energy decreasing, the growth has been increasing significantly and was projected to grow rapidly, particularly if subsidies continued. In 2011 solar capacity was made up of roughly two-thirds utility plant capacity and one-third residential and commercial, primarily rooftop. Approximately 75 percent of the utility capacity was solar cell (PV) and the remainder solar thermal (CSP). Solar cell residential installers were encouraging growth by overcoming the high investment cost to homeowners by offering long-term monthly lease payments that were less than the homeowner's regular electric bill.

California led the country with a third of the capacity, followed by New Jersey (supported by a strong state program), Arizona, and Nevada. California had the largest solar facility, the NextEra Solar Electric Generation Station (SEGS), located in the desert near Barstow. It was a thermal solar operation covering fifteen hundred acres with a capacity of 354 megawatts. It began operation in 1985.

NextEra not only owned the most wind capacity in the United States but, with its California plant leading the way, also owned the most solar capacity. Another one of its facilities is the 75-megawatt Martin Next Generation Solar Energy Center, unique in that in a hybrid arrangement it supplements a natural gas operation. The natural gas–fueled generator adjusts rapidly to fluctuations in the solar generation as clouds drift by or a thunderstorm appears, thus balancing the output. This solar thermal facility sits adjacent to the utility's 3,700-megawatt, gas-fired Martin County power plant near

Solar thermal panels direct the sun onto pipes to heat fluid, which provides the steam that drive generators. (Courtesy of the National Renewable Energy Laboratory.)

the east side of Florida's Lake Okeechobee. One reporter described the five-hundred-acre project as "a set of shimmering mirrors and steel pylons stretching as far as the eye can see." However, because of its small scale, the solar addition reduces the plant's emissions by only 1 percent.[21] The solar center cost $476 million, or over $6 million per megawatt of capacity.

NextEra also had a relatively large solar cell plant for the time, its subsidiary Florida Power and Light's 25-megawatt Desoto Center east of Sarasota. It cost $150 million and was "supported" by the federal government's 2009 stimulus program. President Obama was there to speak at its opening in 2010.

Despite its efforts with solar energy, Florida Power and Light made the following statement in a company released document: Energy from the sun is "virtually limitless," but "it's expensive to convert to useable electricity in Florida. . . . While Florida is the sunshine state, we also experience lots of cloud cover . . . as compared to an arid environment like the Mojave Desert. In Florida, we would need almost twice as much space and equipment to generate an equivalent amount of power as in Arizona."[22]

In America's Southwest, solar energy was seen as a potential solution to the growing power demand in cities like Phoenix and Las Vegas and to help California meet its ambitious renewable mandate. Many facilities were in the planning stage and were to be placed in the California, Arizona, and Nevada deserts with the anticipation that on hot, sunny afternoons, when the demand was highest and operating at full capacity, they could provide as much energy as nuclear reactors. And these solar plants, like wind farms, could be built quickly once approved. The Solar Power Association prepared a list of more than one hundred solar projects, announced but not under way, that would have the potential to add over 10,000 megawatts of solar capacity by 2016. The largest fifty of these ranged in capacity from 550 megawatts to 75 megawatts. A quarter would be solar cell plants, the rest solar thermal. The largest announced plant was the solar thermal Mojave Solar Park, covering nine square miles of desert with an announced cost of $2 billion. It had an advantage in that it could tie to an existing transmission line that was used for an abandoned Mojave coal-fired plant. Due to be operational in 2013, it had faced continuing delays.[23]

Solar projects, like wind projects, faced problems and risks: construction delays and transmission access, permits and government-approval hurdles, federal investment tax credits and state property tax exemptions, sufficient third-party financing, availability of products from suppliers, and the actual energy consumption rate. Add to those cost and environmentally sensitive

This solar cell (photovoltaic) installation is located at Denver International Airport. (Courtesy of the National Renewable Energy Laboratory.)

lands. Such were the factors that were faced as utility-scale solar energy attempted to break out from its small base and match wind energy's growth. California's senator Dianne Feinstein has proposed legislation to convert half a million acres of the Mojave Desert into a national monument, which would impact 10 percent of the projects planned. This illustrates the conflict between those who want to protect the land and those who fear climate change. A federal assessment that determined three thousand tortoises would be disturbed, for example, shut down two-thirds of a project already under construction. (The U.S. Fish and Wildlife in a review declared that the tortoises would not be disturbed and the project proceeded). Meanwhile, the Department of Energy approved $4.5 billion in "conditional" loan guarantees for three desert projects. The department did not specify, however, what "conditional" meant. The loan program had already approved over $38 billion for forty clean projects.[24]

In Arizona, the sun shines at least three hundred days a year, and the state had a mandate to generate 15 percent of its electricity from renewables by

Alternative Sources

2025. Arizona Public Service took the first step in achieving this goal by announcing in 2008 the $1 billion, 280-megawatt Solana Generating Station to be built on alfalfa farmland between Phoenix and the Mexican border. The utility's partner was Abengoa Solar, a Spanish company. In 2010 Abengoa received a conditional $1.45 billion government loan guarantee based on the completion of tax equity financing. Construction had to start by the end of 2010 in order to qualify for the Treasury Department's investment tax credit program. Heated liquid would produce steam for two 140-megawatt generators. Molten salt would store heat, allowing electricity to be generated for six hours after sunset during busy evening use. Natural gas generation was the backup to meet peak demand. Arizona would pay 14 cents per kilowatt hour for the power, with the cost "blended" into the existing nine-cent-per-kilowatt customer rate. The utility's president, Don Brandt, said that the price could be held because of inexpensive energy from the utility's Palo Verde Nuclear Generating Station.[25] The nuclear plant's operating cost was two to three cents per kilowatt hour. As the largest nuclear facility in the United States, it has ten times more capacity than the solar plant. Yet when capacity factor is considered, Palo Verde generates at least forty times more power. Both sources are, of course, clean power.

By 2012 three solar cell plants worldwide were in operation with capacities of 100 megawatts or more, the one in India at 600 megawatts, one in China at 200 megawatts, and one in Ukraine at 100 megawatts. Canada had the 97-megawatt Sarnia plant north of Detroit. In the United States, two new plants exceeded the Florida Power and Light's Desoto installation. Copper Mountain near Boulder City, Nevada, had 48 megawatts of capacity. Its cost of $141 million was well subsidized, with $42 million in federal tax credits and $12 million in state subsidies. As the *Las Vegas Sun* pointed out, it had created only five permanent jobs and all the electricity was being transmitted out of state to California. However, Boulder City will receive $60 million in lease payments (over twenty years) since the plant is on city property. The project has the advantage of being close to Hoover Dam's grid system. The other plant is the 30-megawatt Cimarron Solar Facility in northern New Mexico. The investors are cable TV pioneer Ted Turner and Southern Company; the plant is located on a piece of Turner's vast land holdings in the state.

Many larger solar plants in the Southwest were in development but facing delays. The Treasury Department approved a $646 million loan guarantee by the Energy Department for a 230-megawatt Exelon solar project in Antelope Valley. The project met resistance from local residences and the Los

Angeles Board of Supervisors but was expected to be complete in 2013. It caused controversy because it was so subsidized by the federal and state government, according to the *New York Times,* that the owner, the giant utility Exelon, will be reimbursed for all its upfront investment by 2015.[26]

Solar subsidies were called into question in light of the Obama administration's 2009 stimulus plan, which heavily supported renewables. The administration approved a $325 million loan guarantee to Solyndra, a Colorado company that had developed a new type of solar panel. President Obama was present at the plant's opening. The company was the "poster child" for the administration's support of solar energy. But Chinese low-cost competition ruined the market for Solyndra's panels, and the company went bankrupt in 2011. Its subsidy approval had been rushed initially and early warnings of its problems had been ignored, according to a Republican-led congressional inquiry.

With the exception of NextEra's 354-megawatt California operation, solar plants in the United States were still minimal in capacity by 2012 compared to wind farms. However, this was expected to change when some of the large California solar thermal projects were approved and came on line. With their capacity factors as much as half that of wind, large-capacity plants would be needed to make a significant change in renewable energy generation. Despite an impressive growth year in 2011, solar sources were still less than one-tenth of 1 percent of the United States' electrical generation, compared to wind's 2 percent.

OTHER RENEWABLES

Solar and wind power receive most of the attention, but what about the other renewables—hydro, biomass, and geothermal? With solar and wind, they total 143,000 megawatts, an impressive number representing 14 percent of the United States' overall energy capacity. The leader is hydro, with significantly more capacity than wind. Biomass and geothermal follow, both exceeding solar (see Table 7.1).

Hydropower, though barely acknowledged considering its importance, produces 20 percent of the world's electricity. China, with 179,000 megawatts of capacity, leads the way in hydropower, followed by Canada, the United States, and Brazil, each in the 70,000- to 80,000-megawatt range. The Three Gorges Dam in China is the world's largest power-generation facility at 22,500 megawatts of capacity. The largest in the United States is

Alternative Sources

Table 7.1. U.S. Renewable Energy Installed Capacities by Source, 2011

Source	Capacity (GW)	Percentage
Hydropower	78.7	7.5
Wind	46.0	4.4
Biomass and wood	13.2	1.2
Geothermal	3.5	0.4
Solar	1.6	0.2
Total	**143.0**	**13.7**

Source: U.S. Energy Information Administration, Electric Power Annual Data for 2011, January 30, 2012. Greenchipstocks.com.

the Columbia River's Grand Coulee Dam at 7,000 megawatts. Other large hydro plants in this country, such as the Hoover Dam, are generally in the 2,000-megawatt range. Hydropower, like nuclear power, does not emit carbon dioxide, requires significant construction time and capital, has a low operating cost, and produces reliably at a high capacity factor except when affected by low water levels. However, it demands significant land, displaces people, and affects the environment of waterways where it is placed. Completed in 1985, the Bath County Dam on the Virginia–West Virginia border was the last major hydro plant constructed in the United States.

Biomass, classified as renewable energy derived from wood waste, garbage, and landfill methane (biogas), is in the same "clean" category as wind, solar, and hydro. Yet when burned to generate electricity, it emits carbon dioxide. Because this fuel comes largely from forests that over the long term renew themselves and absorb carbon dioxide, it is considered carbon neutral, thus not affecting global warming. Fossil fuels, on the other hand, release emissions that are not countered by carbon sequestration. While biomass energy is useful in that it does away with waste while generating electricity efficiently (at a high capacity factor), calling it carbon neutral raises questions. In the United States there are about one hundred waste-to-energy plants. The largest is the 140-megawatt New Hope facility in southern Florida, which burns locally grown sugarcane residue and other wastes.

Alternative Sources

Geothermal energy generation also exceeds that of solar. The Geysers, a thirty-square-mile geothermal field ninety miles north of San Francisco, at 1.5 megawatts, contains half the country's geothermal capacity. Its twenty-two power plants provide an estimated 60 percent of the electricity needed in that coastal region north to the Oregon border. It is reliable, predictable, and environmentally friendly and has a high capacity factor. However, it takes considerable permit and construction time to bring on line, and its availability is limited to areas where geothermal reservoirs exist. Deep wells provide dry steam or hot water converted to steam to drive electrical generators. The hot water can also be used to directly heat homes.

COSTLY ALTERNATIVES

Costs of different energy sources are bandied about and are difficult to evaluate and compare. However, the Energy Information Administration provides data that in general clarifies the situation (see Tables 7.2 and 7.3). Subsides are not included in its data. Table 7.2 compares the capital costs of the renew-

Table 7.2. Estimates of Power Plant Capital Costs, No Subsidies Included				
Plant Type Resource	Nominal Plant Capacity (MW)	Capacity Factor (%)	Capital Cost ($ million/MW)	Factored Capital Cost[a] ($ million/MW)
Natural gas	500	87	1.0–2.0	1.1–2.3
Coal	1,300	85	2.8–5.3	3.3–6.2
Nuclear	2,200	90	5.3	5.9
Wind, onshore	100	34	2.4	7.1
Wind, offshore	400	34	6.0	17.6
Solar PV	150	25	6.0	24.0
Solar thermal	100	18	4.7	26.1

[a] Capital Cost adjusted for capacity factor.
Source: Updated Capital Cost Estimates for Electricity Generation Plants, November 2010; Annual Energy Outlook 2011, December 2010. Energy Information Administration.

ables, wind and solar energy, to the base load sources, coal, nuclear power, and natural gas, which has recently been added to that category. For each source, typical (nominal) plant size and capacity factor are shown in the table. Natural gas has the lowest capital cost per megawatt, followed by coal and wind, which are comparable. Nuclear power, offshore wind, and solar energy are much higher. When capital cost is adjusted by capacity factor, natural gas becomes the cheapest source by far and nuclear power moves in ahead of wind. Solar power and offshore wind become four times more expensive.

The Energy Information Administration has estimated in cents per kilowatt-hour the cost of new plants going into service in 2016 (see Table 7.3). The total system cost, including capital and operations, shows natural gas continuing to be the least expensive, followed by coal, wind, and nuclear power, in that order. Solar power and offshore wind are, again, far more expensive. When carbon capture is included in the overall cost of coal, wind becomes cheaper than coal.

Similar Energy Information Administration data for other renewables shows geothermal and biomass having capacity factors in the same range as

Table 7.3. Estimated Cost of New Generation Plants Entering Service in 2016, No Subsidies Included			
Plant Type	**Capital Cost (cents per kWh)**	**Operations Cost (cents per kW/h)**	**System Cost (cents per kWh)**
Natural Gas	1.8–3.5	4.8–5.4	6.6–8.9
Coal	6.5–9.3	3.0–4.3	9.5–13.6
Wind	8.4	1.3	9.7
Nuclear	9.0	2.4	11.4
Solar PV	19.5	1.9	21.1
Wind—Offshore	20.9	3.4	24.3
Solar Thermal	25.9	5.2	31.1

Source: Energy Information Administration, "Annual Energy Outlook 2011."

nuclear power and the fossil fuels. Hydro's factor is half that. With regard to system cost, hydropower is a little less than coal and wind. Geothermal and biomass costs are similar to nuclear power.

With wind approaching cost parity with coal, does this mean that it is at a point where it can stand on its own? Can it survive without government subsidies? Can wind compete with natural gas's low cost and high capacity? Over the past few years, wind has outperformed all other energy sources in its growth rate, but there are ongoing problems ahead that could slow it down.

Chapter 8

Growth Issues

TURBINES AND PEOPLE

Wind energy has grown rapidly despite economic and environmental obstacles. Environmentally, bird killings in the California passes, the impact on bats on the Appalachian ridgelines, and the effect on nesting turtles on the Florida beaches were detriments. A North Dakota wind farm was canceled due to the fear of killing hawks. Economic concerns were raised about the impact on nearby real estate values and tourism in popular nature areas. Some of these issues became important in siting wind farms; in other cases, they didn't matter.

But there were issues that did matter and could become more serious in slowing wind energy's future growth. More turbines, and taking more land, pressure more people. The nation needs transmission lines to absorb, manage, and keep stable the variations in the flow of increasing amounts of wind energy. And the country requires a stable energy policy with continuing subsidies for wind energy.

The Department of Energy's report on 20 percent wind energy by 2030, *Wind Power in America's Future,* recognized the NIMBY situation by stating that wind projects "generally enjoy broad public support" but can "raise concerns in local communities."[1] The report stated that local concerns caused about 10 to 25 percent of proposed projects not to be built or to be significantly delayed. It noted that the best areas for turbines were those with the least population density, and yet those areas "may be prized for tranquility, open space, and expansive vistas." This was certainly true with the Beech Ridge project in West Virginia, and it is an obstacle to expansion in areas like

New York State, where the space between people and turbines is limited and where some strong feelings have been expressed.

Invenergy developed the 112-megawatt High Sheldon Energy Wind Farm located twenty-five miles outside of Buffalo, which went on line in 2009. Sheldon councilman Glenn Cramer has expressed regrets over his participation in activities related to High Sheldon. In response to a nearby landowner who was trying to sell his town on a new wind farm, Cramer posted some comments comparing the relationship with the wind developer "to a relationship with the Devil himself." The town, he said, was "hopelessly divided." The wind farm destroyed TV and radio reception in many areas, and Buffalo radar couldn't forecast weather conditions accurately east of the farm. A home in the area sold with turbine concerns—"rattling windows, shadow flicker, and an unending thump, thumping"—for less than the owner wanted. The owner wrote that Sheldon was "another example of why industrial wind farms do not belong anywhere near people. . . . Some residents have gained from the wind farm, but it has been at the expense of their neighbors."[2]

A writer in the *Watertown Daily Times* said that he could see 151 turbines of the Maple Ridge Wind Farm out his front door on the Tug Hill Plateau. It had "fragmented wildlife habitat, the community, and our family." TV and radio were affected; he couldn't get cable TV. He and his family were in the middle of "a major industrial nightmare." His taxes hadn't been reduced, though the county received $9 million annually from the developer, which in turn received a major property tax reimbursement from the state. The heavy equipment used to repair the broken turbines destroyed the streets. "It's all about money for the developers—the turbines (Vestas) are foreign owned—so they can take our tax money back to their country," said Yancey.[3]

Safety concerns became an issue after the collapse of the turbine towers at the Altoona and Fenner wind farms in 2009. Shortly after the Altoona incident, the nearby town council of Beekmantown killed a wind project that had been pending for three years. It changed Councilman Samuel Dyer's mind. "Excuse my language, but that thing in Altoona scared the s— out of me," he said.[4]

In 2006, Invenergy, led by Dave Groberg, who was handling the Beach Ridge project at the same time, proposed a wind farm in Delaware County on the western slopes of the Catskill Mountains, one hundred miles from New York City. This is an area of wooded hills, green valleys, fishing streams, ski slopes, small farms, and historic, quaint villages, very similar to Greenbrier County. Invenergy wanted to place thirty-three turbines within the townships of Stamford and Roxbury along eight miles of the Mooresville mountain

A group of wind turbines surround a house on the Tug Hill plateau in western New York State. (Courtesy of National Wind Watch.)

range. The Stamford Town Council received 820 letters from 20 percent of the local taxpayers opposed to "industrial wind turbines." Councilwoman Katherine Egret, a longtime member of the council, said, "I hate to see all the stress in the community it has caused. The majority doesn't seem to want them."[5] The town passed an ordinance with restrictions and regulations that required Invenergy to submit an extensive application. The state then had to review the project, a nine- to twelve-month process. Only then could the Stamford Town Council rule on the project.

Mike Triolo, a Stamford Town Board member, had voted for the ordinance. In some remarks at the time, he took a broader stance. He said, "Some of you will disagree with tonight's decision but nobody is going to be 100% happy with any wind power ordinance." He talked about researching every item to the best of his ability, and "given the amount of misinformation . . . generated by both opponents and proponents, that was not an easy task. . . . Being green is not easy. It requires conservation, some discomfort and, yes, maybe having obtrusive structures somewhere in the vicinity." He stated that wind energy was not the answer, but it was one of many answers, including solar, clean coal, hydro, biomass, natural gas, and even nuclear energy. For those who feared that turbines would change the character of the region, Triolo said that the region had already changed and been irreparably damaged, with farms subdivided, houses on hillsides, dusk-to-dawn lights, animal habitats destroyed, and so on. "Where man enters, damage is created. Some call it progress, some ecological disaster, and some accept it as a consequence of life. It is incumbent on us to balance the effects of those changes." He concluded his remarks by saying, "I've been waiting over thirty years for a rational energy policy in this country. Obviously Washington is incapable of handling the issue and it must start somewhere."[6]

But eighty-one-year-old Floyd Many, a member of the board for twenty-five years, didn't see it Triolo's way. He said, "I will say this just once: not in my backyard. People in Delaware County think it ought to be in the Adirondacks. People in the Adirondacks think it should be in the ocean off Massachusetts. Teddy Kennedy thinks it should be someplace else. Everyone wants alternative energy, but no one wants it where they have to look at it."[7] Five years later, construction of this wind farm had not started.

And, of course, Florida Power and Light had searched three hundred miles of the state's eastern seaboard seeking an undeveloped area for a small wind farm. They hadn't succeeded, even with a reduced number of turbines. People were in the way.

The major issue in meeting the Department of Energy's scenario of 20 percent wind, as it admitted in its report, was transmission lines. They were needed to access new and remote generation, to relieve congestion on the lines, and to improve system reliability. The department described the need for adding twelve thousand miles of transmission lines to the two hundred thousand miles already operating at a net present value cost of $20 billion. And this was probably understated. Utility and transmission giant American Electric Power thought in terms of nineteen thousand miles of new lines at a cost of $26 billion.[8] In the West, new transmission lines would extend south and west, feeding the Seattle, San Francisco, and Los Angeles areas. In the Midwest, new lines would run from almost every state eastward. In the East, with a thick network already in existence, more short interconnects were needed.

If wind energy could be stored, and then fed into the grid as needed, then the fluctuations in its volume could be controlled. In the Midwest of the 1920s, farm windmills worked because the small amount of power generated could be fed into a storage battery and used when the farmer turned on a light. But no such battery or storage system had been developed. Thousands of megawatts generated by wind farms needed to be stored on windy days and dispersed to meet peak needs on calm days. Excel Energy, a Minnesota-based utility with operations in eight midwestern and western states, was evaluating a wind to battery system. The 1-megawatt test system weighed eighty tons and was the size of two tractor-trailers, a far cry from the farm windmill battery storage. However, this system could store 7,200 kilowatt-hours of electricity, enough to power five hundred homes for seven hours. Excel was working with the National Renewable Energy Laboratory on the project.[9] While this project was in its infancy, it was a beginning.

Also in Minnesota, the Public Utilities Commission had the foresight to do a study showing that it was feasible to integrate significant amounts of wind energy into its power grid. The state could increase installed wind capacity to 6,000 megawatts with a capacity factor of 40 percent and integrate it into its grid, thus meeting the legislature's renewable energy standard. The concern was not so much the availability of the wind but its variability, the minute-by-minute fluctuations in the electricity produced. It was important that sites were dispersed so that there would be a smoothing of wind variations, thus providing a stable supply. At the time most of the wind farms were located across the southwest corner of the state on elevations up to two thousand feet. Still, backup power was needed, particularly from natural gas power plants, to compensate for the changes in wind generation. Transmission line improvements would have to be incorporated into the power system to provide multiple points of entry for the wind power. And the four utilities supplying power to the state would have to act as one "balancing authority" in controlling the grid.[10]

If wind energy was to grow, the issue really came down to transmission lines. In the East, wind farms were generally located near transmission lines. The Beech Ridge project in West Virginia, despite being in the mountains, had to run just a thirteen-mile line to connect itself to the grid. A wind farm in northern West Virginia was located within sight of a huge coal plant and its existing transmission system, which fed eastern metropolitan areas. The Florida beachside project, if it had happened, would have used the existing nuclear plant's lines. With limited capacity in the East, wind wasn't yet generating sufficient power to overload transmission lines to any degree.

Wind energy growth in the Northwest benefited from under-used transmission lines associated with the giant Grand Coulee Dam system. Power agencies did have concerns of overloading the grid in the spring, when both the hydro and the wind systems would be generating maximum power as the snowmelt loaded the dams and the seasonal winds spun the turbines. The remoteness of the large Shepherds Flat wind farm in north-central Oregon was solved by the Bonneville Power Authority running a connection line to its grid. Transmission lines existed south to California. But with further wind energy growth in the area, would they soon become congested?

In California, as the state scrambled to meet renewable energy goals, the remoteness of wind farms was a determent. In order for the small, initial phase of the ambitious 3,000-megawatt Alta wind project to begin, it required a $1.8 billion transmission line that was to be completed in 2013. This was one of three transmission projects underway out of eleven that would be necessary to meet California's 33 percent renewable goal. It would take fourteen years and cost $16 billion in a state that was experiencing both siting opposition and financial difficulties.

Transmission was the biggest problem Texas and the Midwest faced. The problem varied from state to state, depending on the remoteness of wind sites, the availability of urban markets, and the difficulties, political and financial, of installing new transmission lines.

The continuing expansion of wind farms in Texas caused a critical need for transmission lines. Even with half the wind capacity at the time com-

Transmission lines like theses must be constructed across the country to bring wind and solar energy to high population markets if renewable energy is to grow significantly. (Courtesy of the National Renewable Energy Laboratory.)

pared to 2010, turbines were being shut down when the wind blew because the grid couldn't handle the flow, causing excess wind power to be dumped. In 2008, the Electric Reliability Council of Texas (ERCOT) decided to build a $4.93 billion transmission system to bring wind energy from West Texas to the metropolitan areas of Dallas, Houston, Austin, and San Antonio. ERCOT managed the distribution of electricity for 80 percent of the state's power. It was unique in that being totally within the state, it was not subject to the

Federal Energy Regulatory Commission as were other U.S. grids. The new line would be in operation in 2013.

T. Boone Pickens had canceled his wind energy plans due to the cost of a transmission line, but NextEra, not waiting for ERCOT, built its own line, stretching from its huge Horse Hollow wind farm to a substation near San Antonio. The line crossed Texas's scenic Hill Country, where, unlike the turbines in the wide-open spaces of West Texas, it met opposition.

By January 2009, the Electric Reliability Council of Texas had let contracts for its transmission system. Grid congestion had worsened in 2008, but with more wind turbines coming on line, consumption in Texas from wind energy had almost doubled in a year. The Texas Public Utility System stated that the new lines would increase monthly power bills three to six dollars per month for a period that could run anywhere from five to fifteen years.

On Sunday, February 28, 2010, during the noon hour, with strong winds blowing across West Texas, electricity generated by wind turbines momentarily peaked, providing 22 percent of the power consumed in the state. Wind energy generation was operating briefly at a capacity factor of 70 percent. Yet because of grid limitations, wind energy had to be curtailed because the supply exceeded what the transmission lines could handle.[11] (This had occurred in Denmark and Germany, where generation at times exceeded the grid's ability to handle it.) With ERCOT's new transmission line in operation, this situation presumably would not have occurred, and even more wind energy could have been sent to Dallas and Fort Worth.

Iowa, which was second only to Texas in wind capacity, led all states in the percentage of power generated by wind. Yet wind energy authorities were concerned about future growth. Iowa was now at a point where it was "hitting a wall." Unlike Texas, which had its own transmission system, Iowa had to deal with other states in making transmission decisions. By 2010 there was no growth in the state's wind capacity, but it picked up again in 2011. How much of the slowdown was caused by "hitting the wall" or by other problems was not clear. Nearby states such as North Dakota, Minnesota, Illinois, and Indiana had shown some growth. North Dakota and its huge Hartland Energy Wind Center were waiting on Congress to act to provide a solution to its transmission gridlock. The state, unlike Minnesota with Minneapolis, had no urban markets. It had vast wind resources that if developed would have to be exported to other states.

A consortium of six grid operators in a 2009 study estimated that to support the country receiving 20 percent of its electricity from wind would require $80 billion in transmission lines in the eastern half of the United

Growth Issues

States. This cost was far higher than that estimated by the Department of Energy or American Electric Power.

But cost is not the only concern with transmission lines. As with wind turbines, some people don't like transmission line towers. Regional grid operator PJM Interconnection wanted the 275-mile Potomac-Appalachian Transmission Highline (PATH) built across West Virginia, Maryland, and Virginia to improve system performance. Effort was suspended because of objections and demands that other alternatives be reviewed. The line would have transmitted power from coal-fired plants eastward. A similar situation occurred in California. A six-hundred-mile, $1.5 billion line was planned to transmit power from northern California to Sacramento. Communities along its path fought it, and the three major utilities involved backed out of the deal, canceling the project. It was specifically planned to transmit wind, solar, and geothermal energy.

TAX BENEFITS ARE SO IMPORTANT

If anything was going to make or break wind energy development, it was going to be the government. State governments in California, Texas, and Iowa had promoted wind with renewable energy standards and incentives, each to a different degree. The federal government was a different matter, however. Its production tax credits, of all the subsidies, were the most effective in the growth of wind energy. With the 2.2-cent-per-kilowatt-hour tax write-off on every kilowatt generated, they made the cost of wind competitive and investment in wind by developers like NextEra and Invenergy profitable. (The credit was good for the first ten years of operation.) Without the production tax credit, there would be few wind farms. And yet the tax credits had been, like the wind, variable. Congress had extended them on a year-to-year basis, but with three lapses—1999, 2001, and 2003—that all but stopped wind farm construction in each of the following years. However, with the stimulus bill, the production tax credit was extended through 2012. (As a result of the fiscal deal between Congress and President Obama at the end of 2012, the production tax credit was extended one more year for wind energy projects starting before the end of 2013.)

Congress passed President Obama's $787 billion stimulus package, known as the American Recovery and Reinvestment Act, on February 17, 2009. Depending on who was counting, since the list of items funded seemed infinite, one source claimed as much as $60 billion for "green initiatives" and $20 billion for "green tax incentives." The bill came right in the midst of the great recession, as the credit crisis struck and investment dried up, thus presumably giving a boost to wind energy. It included a 30 percent investment tax credit for developers but was limited to construction starting in 2009 and 2010 and covered only turbines and other associated equipment, not roads and transmission lines. The stimulus bill even allowed a developer to convert the investment tax credit into a 30 percent cash grant if accepted by the government. There was an immediate 50 percent depreciation for the capital cost of wind equipment acquired in 2009, bonds to finance facilities generating electricity from clean renewable energy such as wind, and a 30 percent tax credit for energy manufacturing projects producing equipment for renewable resources. Guaranteed loans were available for a variety of projects and technologies if construction started by September 2011. These included transmission and interconnection improvements, smart grid planning, and so on.

In addition to wind and solar energy, funds were allocated for home weatherization, efficient appliances, advanced batteries, and transportation electrification. Fossil fuel received $3.4 billion for research and development covering low-emission coal plants and carbon capture projects. Nuclear energy received $6 billion for cleanup and decontamination, particularly for nuclear defense sites.[12]

As an individual at the American Wind Energy Association 2009 fall symposium in Orlando in an understatement said, "Tax benefits are so important." And so they are. Congress asked the Energy Information Administration to study the energy subsidies for 2007. They totaled almost $17 billion, of which renewables received $4.9 billion, almost double any other energy source. Yet in that year wind generated less than 1 percent of the nation's energy. The subsidy for wind was twenty-three cents per kilowatt hour when applying the portion that directly subsidized electricity production, far higher than any other energy source except "refined coal" (recall the synfuel scam, as some people called it, soon canceled by Congress).

The House of Representatives, then controlled by Democrats, quickly followed the stimulus bill with the first climate change legislation. The Waxman-Markey, thirteen-hundred-page, cap and trade bill passed by a narrow vote. The bill's goal was a 17 percent reduction in carbon dioxide emissions from the 2005 level by 2020 and 83 percent by 2050. There seemed to be no explanation as to why 17 percent and 83 percent were chosen. The bill set a national standard of 20 percent power generation from renewable energy by 2020. One-third of that could come from energy efficiency (conservation) rather than renewable sources. Billions of more dollars would be provided for energy projects, low-carbon agricultural programs, clean coal research, and the development of electric vehicles. The bill "grew fat with compromises, carve outs, concessions, and out and out gifts intended to win the votes" of lawmakers and the support of industries, said the *New York Times*. White House Chief of Staff Rahm Emanuel expressed President Obama's feelings about the bill: "He loves the bill and lobbied hard for it, including the great, the good, and the not-so-good provisions."[13]

This patchwork bill contained something for everyone except for the consumer, whose household cost of living would increase $175 a year, according to the Congressional Budget Office. Democratic representative John Dingell was caught saying, "Nobody in this country realizes that the cap and trade is a tax, and it's a great big one."[14] There was also the opinion that a 17 percent emission reduction would be an inconsequential amount on a global basis. The Senate never considered the House bill.

But the political activity affecting renewable energy in 2009 was hardly over. In the fall of 2009, the discovery of e-mails from Britain's East Anglia University's Climate Research Unit raised questions about the 2008 United Nations Intergovernmental Panel on Climate Change report. The disclosure exposed the possibility that climate data had been exaggerated to promote the theory that man was the major cause of climate change. For example, the report stated that the Himalayan glaciers would melt and disappear by 2035. Rajendra Pachauri, who led the UN panel and shared the Nobel Prize with Al Gore in 2007, was criticized for scientific sloppiness. *Time* magazine indicated that in reality there was little in the e-mails to discredit the temperature data and warned that "science itself risks becoming a political debate, like everything else today, with no room for objective data or authority."[15] The Pew Research Center indicated that Americans' belief in warming had dropped from 71 percent in April 2008 to 57 percent in October 2009.

Despite this environment, the Senate began to act on a climate bill. Six committees were involved, but Republicans were reluctant. To expedite matters, Democrat John Kerry, independent Joe Lieberman, and Republican Lindsey Graham formed a coalition to put together what they called the American Power Act. Their middle-ground approach was different from that of the House in two major ways: more support for nuclear energy and the expansion of offshore drilling. They kept the House goals of reducing emissions by 17 percent below 2005 levels by 2020 and 80 percent by 2050. Their intent was to reduce business costs and increase the nation's production of oil, gas, and nuclear power. But Lindsey Graham abandoned the effort just as it was to be announced. Graham, who had taken a major political risk in supporting the bill, was angry that in a political move Harry Reid, the Senate majority leader, had placed immigration legislation ahead of the climate change bill. Congressional action on climate change was dead with no prospects of being revived in the foreseeable future, particularly when a Republican-majority House was elected in 2010.

At the state level, as finances became tighter, renewable energy raised questions. The Oregon legislature had concerns about subsidies for wind energy developers. Over an eight-year period the state had accepted applications for $300 million in tax breaks. The legislature wanted to cut by two-thirds the level of expenditures, but Governor Ted Kulongoski vetoed the bill, fearing that it would slow the growth of wind energy in the state. Portland's newspaper, the *Oregonian,* carried out a study of the effect of the state's Business Energy Tax Credit and determined it was not a major factor in developers' decisions to build wind facilities, even though it was one of the most "lavish" given out by a state. The credit was 50 percent of the first $20 million of a wind developer's cost. On a small development this was significant, but on a large, 100-megawatt project, it was not. The criticism of the credit was that the money could be better spent on schools and other state needs rather than on wind developers, which already were receiving substantial federal and local subsidies. Counties were allowed to limit assessment to the first $25 million of a project's investment cost. One county assessor, Linda Hill, said of the developers, "They will always complain unless they are paying nothing. You wouldn't see all those wind farms going up if it wasn't hugely profitable."[16]

In Congress, with no consensus between the House and the Senate on climate change, the UN report being questioned, state concerns increasing, and the public's faith in the issue declining, President Obama attended the international climate meeting in Copenhagen in December 2009. Using the Waxman-Markey bill goal, he bravely declared that the United States intended to reduce greenhouse gas emissions by 17 percent below 2005 by 2020. (In 1997 at Kyoto, President Clinton agreed to a 5 percent reduction in emissions, but Congress didn't ratify the treaty.) The result of this contentious Copenhagen meeting of 190 countries was an "accord," a document described as not formally adopted, not binding, and not able to set a deadline for a climate change treaty. During the conference, countries made promises or pledges like those President Obama made for the United States. But even if all the promises were met, analysts felt that the main goal—to limit global warming to a 3.6 degree Fahrenheit rise since the preindustrial era—could not be met.

Would wind energy, with all its recent growth, begin to reach its limit? With government backing, it had certainly outpaced other energy sources in its rate of growth. In reality, could it stand on its own and continue to be the energy most noted in the drive to reduce greenhouse gas emissions?

The Beech Ridge operation, like many wind projects, had advanced despite the viewshed opposition. But another problem surfaced that Invenergy assumed had already been resolved.

Chapter 9

Endangered

Construction on the Greenbrier wind energy site began, but as John Stroud had said, "This is not the end." On June 10, 2009, Mountain Communities and the Animal Welfare Institute (AWI), a fifty-year-old animal protection organization based in Washington, D.C., filed a lawsuit against Beech Ridge in federal district court in Maryland. The suit stated that Beech Ridge should be required to obtain, under the Endangered Species Act, a federal permit from the U.S. Fish and Wildlife Service "if death or harm could come to an endangered species." The death or harm came from the spinning blades; the endangered species was the Indiana bat, which lived in and migrated through the area. The Washington lawyer filing the suit, Bill Eubanks, said, "We're not asking for a permanent halting to the project. Construction should be halted until Beech Ridge gets the permit. . . . We don't want to see the Indiana bat die off."[1]

Invenergy's general counsel said the lawsuit lacked merit. One month later, the AWI and Mountain Communities added to the lawsuit by filing for a preliminary injunction. Because Invenergy had moved forward rapidly with construction after the filing of the complaint, AWI and Mountain Communities felt they had no alternative but to file the motion. Some believed this was the first lawsuit that challenged a wind energy project on environmental grounds. The suit stated that other "poorly sited" wind power projects had already destroyed and maimed countless numbers of bats.

Dave Cowan, who lived near the site, had been asked to be a plaintiff in the lawsuit too because he "knew bats." Shortly after the lawsuit was submitted, he reported to the sheriff in Lewisburg that his yellow mailbox, which bore images of bats and owls, had been vandalized. "This seems to be more than coincidence and not a case of teenage vandalism," he said.[2] He lived in an area on the western side of the wind site, where most people favored the project.

INDIANA BAT

The Industrial Wind Action Group, which noted on its website that it provided the "exposure of wind energy's real impacts," provided information in an editorial on bats and the Beech Ridge site. A "world-renowned expert" on bats, Thomas H. Kuhn, forecasted in a paper that by 2020, 111,000 bats could be killed in the mid-Atlantic highlands by wind turbines. This concern over bats began in 2003, when significant kills were reported at NextEra Energy's Mountaineer Energy center in Tucker County, West Virginia, on forested ridgetops similar to those in Greenbrier County. BHE Environmental, which had done a study for Beech Ridge, even predicted 135,000 kills during the twenty-year life of the project. Wind Action further stated that hundreds of bats reside within ten miles of the project, and some are as close as seven miles from turbines. Indiana bats live to the south and east of the project and forage and roost during the summer to the west and north. BHE and Beech Ridge took the position that no Indiana bats were likely to be killed on the site because none had been found there during the July 2005 study. However, it was discovered in pre-trial investigation for the federal suit that ultrasound and acoustic data had been taken at that time but ended up in a file cabinet and had never been analyzed. Two experts reviewed the data for the plaintiffs and determined that Indiana bats "were almost certainly present" during the 2005 study.[3]

While men in lumber shirts, dungarees, and hard hats worked the site, men in dark suits and ties were again discussing and making judgments on the project's fate, this time in a federal courtroom in Greenbelt, Maryland.

On the first day in court, the *Washington Post* ran a front-page article on the suit. Dave Cowan, the Greenbrier County "bat expert," was shown in a photo standing in front of the completed turbine on the Beech Ridge site. He was portrayed as "a long time caving fanatic who grew to love bats as he slithered through tunnels from Maine to Miami." The Indiana bat that he was in court to protect was described as a "creature that weighs about as much as three pennies and, wings outstretched, measures about eight inches." Mountain Communities and AWI's lawyer, Eric R. Glitzenstein, said in his opening remarks that both sides agreed 130,000 bats of all types would be killed over

Endangered

the next twenty years. He said that Beech Ridge's position was "let's roll the dice and see what happens." He continued: "This approach to the Endangered Species Act is not in keeping with what Congress had in mind." Beech Ridge's lawyer, Clifford J. Zatz, specializing in environmental and toxic tort litigation, responded: "A $300 million, environmental friendly, clean, renewable energy project waiting to serve 50,000 households is in limbo over a rare bat nobody has ever seen on the project site."[4]

Beech Ridge's consultant had put up nets at the site during the summers of 2005 and 2006, and no Indiana bats were caught. However, the bats migrate not in the summer but in the fall. The Indiana bat's growth rate was already limited by the female having only one baby a year, and now the bat was also subject to a disease that was reducing its numbers. Not only did the turbine blades kill the bats, but in a phenomena called barotraumas, the spinning blades create a low-pressure area that causes the bats lungs to hemorrhage. The Public Service Commission of West Virginia, as one of its conditions, had addressed the bat issue by requiring Beech Ridge to keep records of bat kills for three years. To mitigate the problem, turbines could be shut down at certain times, thus reducing kills. But Dave Cowan had only one position in mind. He told the *Post*, "I think if the turbines kill one Indiana bat, that ought to end it. They ought to shut it down."

Was the Indiana bat like the snail darter, which halted the operation of an almost completed TVA hydropower dam project in Tennessee from 1973 to 1979? The Endangered Species Act protected the fish, a minnow the length of three small paper clips. Proceeding through a suit to stop the project and subsequent judicial appeals, the case reached the U.S. Supreme Court in 1978. The Court ruled that the Endangered Species Act, as established by Congress, forbids projects that would eliminate species protected by the act. Tennessee senator Howard Baker, on a second try, pushed through the Senate an amendment to the Endangered Species Act that granted an exception for the TVA dam. He pointed out that because of this small fish, a dam that was to supply hydro energy and was almost complete was to be destroyed at a time when the country was facing an energy crisis.

Dave Groberg said that during the hearing the snail darter would be discussed, but he didn't think the precedent fit their case. Mountain Communities' John Stroud said that the snail darter would be brought up. He also said that the final arguments would be completed in two to three days and that the judge was speeding up the process because the project was on hold. A decision was expected within four months, perhaps sooner, much sooner than the normal full year.[5]

125

COURT ORDERS

On December 8, 2009, U.S. District Judge Roger W. Titus ordered Beech Ridge Energy and Invenergy to cease construction on any turbines other than the forty already being built and to not operate any turbines between April 1 and November 15 of any year.[6] Beech Ridge "will violate" the Endangered Species Act in regard to Indiana bats, he declared, unless they obtain an Incidental Take Permit from the U.S. Fish and Wildlife Service. This permit is required when activities threaten an endangered species and must include a habitat conservation plan that "ensures that the effects" are "adequately minimized and mitigated." The only way the Beech Ridge project could continue was with a permit that would indicate locations where turbines were appropriate and where they were not. The turbines could operate during the winter months, when the bats were in hibernation.

Judge Titus, in his written comments, stated that the case was about bats, wind turbines, and two federal policies—the protection of endangered species and encouragement for the development of renewable energy. The Fish and Wildlife Service had designated the Indiana bat years ago as an endangered species. He wrote that it was "uncontroverted that wind turbines kill or injure bats . . . that there is a virtual certainty . . . the Beech Ridge Project will take endangered Indiana bats" in violation of the Endangered Species Act. "The tragedy of the case" was that Beech Ridge disregarded the advice of the Fish and Wildlife Service by not preparing an Incidental Take Permit in advance. The judge chastised both Beech Ridge's environmental consultant and Beech Ridge, who "disregarded" what they were told. The Fish and Wildlife Service wanted three years of radar, acoustical, and thermal surveys to determine bats coming out of local caves after hibernation. In addition, they wanted two years of mist-netting surveys. Beech Ridge's attorney had opposed these surveys because of the financial burden and the construction delays. In essence, the environmental consultant had taken the "minimalist approach," conducting only one type of survey and even that for only one season.

With his order the judge "invited" the parties to resolve the case so that there would be no further legal action beyond Beech Ridge submitting the Incidental Take Permit. On January 19, 2010, the lawyers for Beech Ridge and for the Animal Welfare Institute and Mountain Communities agreed to a stipulation that in part would supersede the original order.[7] Beech Ridge would not construct forty turbines on the eastern side of the project. This included twelve turbines to the south that overlooked the Williamsburg

126

Valley. Some of these would have to be decommissioned because of concrete foundations built in 2009. The remaining twenty-two sites, upon which no work had been done, were to the northeast. In doing this, Beech Ridge would "ameliorate" the potential impact on the Indiana bats. The forty turbines already constructed, which remained on the western side of the project, could now operate not only during the winter hibernation months but also during daylight hours the rest of the year. This decision was based on receipt of the Fish and Wildlife Service's permit. Seventeen additional turbines planned for Phase I and ten for Phase II could go ahead and be built subject to the same conditions as the forty existing units. However, the land clearing and construction at the Phase II sites could occur only during the bat hibernation months. Pending West Virginia Public Service and Fish and Wildlife Service approval, thirty-three additional turbines could be placed in the western areas. This gave Beech Ridge the potential to have a one-hundred-turbine project.

In this agreement, Mountain Communities would play a "constructive, cooperative role" in the ongoing process, would "not challenge" any Fish and Wildlife Service findings and conditions, and would not "pursue or block" project construction or operation. After almost five years and two previous attempts at compromise, the battle between Beech Ridge and Mountain Communities was over. The developer could move on with the project, and the opponents had succeeded in reducing the number of turbines overlooking their homes and farms.

BETTER TO SETTLE

Mountain Communities' John Stroud talked about the project in the summer of 2010 as he sat at a picnic table under the trees by his farmhouse.[8] Behind him was a field full of grazing sheep, and beyond, rising two thousand feet above the field, was the Nunly Mountain ridge. The first of nineteen turbines to be placed on this ridge was to have been above his house, with the rest running toward Cold Knob. However, the federal court decision and the agreement between Beech Ridge and Mountain Communities resulted in ten of the these turbines overlooking Williamsburg and all the turbines to the northeast, which made up the eastern side of the project, being canceled. Foundations had already been laid for ten that had been deleted and would have to be removed. All the deleted turbines, particularly those to the northeast, were the closest to the bats. However, there were still fifteen to sixteen

Endangered

turbines left that could be seen clearly from the approaches to Williamsburg and from Route 219 running north out of Lewisburg, twelve miles away. From a ridge on the east edge of Lewisburg they were barely discernible on a clear day, but at night the blinking red lights on the towers were apparent. From a corner of his yard, John Stroud could see only one turbine, and then only the upper end of the slowly turning blades poking above the ridgeline.

Stroud had attended the federal court hearings. Beech Ridge gave up six turbines before the hearings started, perhaps, as Stroud speculated, because of the bank crisis and the difficulty of getting financing (or was it also a gesture to the court?). The "smoking gun," as he put it, was the bat study done for Beech Ridge. In discovery, Mountain Community's lawyers reviewed thirteen thousand pages of Beech Ridge information and found a memo suggesting the suppression of some of the bat information. The memo said, according to Stroud, "We need to take possession of the recordings, or we are in deep trouble." The recordings indicated the presence of bats. The Washington law firm Meyer, Glitzenstein, and Crystal, specializing in wildlife and animal protection issues and working with nonprofit organizations at "below market rates," represented Mountain Communities. Stroud spoke of Eric Glitzenstein: "He was probably fifty. He is such a good lawyer that it was like watching a TV show. He totally mastered the issues. Eric Glitzenstein knew all the statistical stuff when the other lawyer did not. He had a poker face and never cracked a smile when he was right about something. He would literally tear up what the other was saying in cross examination." About Judge Titus, John said that he was "very fair and very attentive," unlike others during PSC and Supreme Court hearings, who had acted like "they didn't care." The judge cross-examined and reviewed documents to be sure of the issues. He followed the law, though he might not always have agreed with the law. He was quiet and serious. He identified the wind turbines as a business and treated Beech Ridge fairly. Judge Titus said that the two parties should get together and come to an agreement. And that they did, to the extent that Beech Ridge's turbine locations and final number of units would depend on Fish and Wildlife decisions. And there was the possibility that in their studies, Fish and Wildlife might discover even more endangered species than just the bats.

While John Stroud and his neighbors would have preferred all the turbines seen from the Williamsburg Valley to have been removed as well, they accepted the agreement for several reasons. They had won the case and Beech Ridge had to pay the legal fees. This was fortunate since Mountain Communities had financed itself largely through its members' monthly pay-

Partially constructed wind turbine sites line a ridgeline in Greenbrier County, West Virginia. Some of the sites will be decommissioned due to a federal court ruling. (Courtesy of Mountain Communities for Responsible Energy.)

ments and a few local fundraising events. If they had asked for more, Beech Ridge might have appealed. The company's request to add additional areas where terminals could be placed had been denied. However, within the areas where turbines already existed or were planned, they could add more, giving them a potential one hundred for the project. Any appeal would have gone to the Fourth Circuit Court in Richmond, a conservative court with a reputation for ruling in favor of coal companies on environmental issues. Another federal case settled three weeks later had gone against endangered species and could have set an unfavorable precedent. And all this would be taking place in a political environment in which the Obama administration tended to care more for renewable energy than endangered species. If Mountain Communities had lost, they would have to pay legal fees. As Stroud said, he and his allies were "at a dead end and felt that they should settle since they got pretty much what they wanted."

Debbie Sizemore, co-chairperson of Mountain Communities, had said at one time she would never compromise. "My heart was with Debbie," Stroud said. "She didn't want to settle but finally agreed that it might be better if they did."

KNOW THE DETAILS

Looking back on five years of opposition to the wind turbines, Stroud talked about his role in it.[9] He said that before it all happened he had had a nightmare. He dreamt that he walked out on his porch and saw six gigantic smokestacks on the ridge above him. In 2005, when it became clear the turbines were coming, he began doing research on wind energy and reading what opposing organizations and individuals were saying. As a result he knew a lot compared to his neighbors. This led him to become the co-chair with Sizemore when Mountain Communities was formed. He didn't like talking in front of people, so Debbie led the meetings in the beginning. In her job with the state health department, she was used to it. He learned that a person could not talk about wind energy superficially; they had to know the details. "It was not easy being against it," he said. "You almost have to give a lecture." It would be his PowerPoint show that presented Mountain Communities' position against the wind turbines being placed overlooking the farms of the Williamsburg Valley.

Stroud's smokestack dream had made him realize that he was very interested in "aesthetics"; they mattered to him. He liked things to look nice. "I have seen what the turbines have done to Wisconsin and it's tragic," he said. "They don't work and they're almost a scam," he added. "Nobody is willing to listen when they are told that. Some people say when they are driving and the turbines are turning in front of them, it gives then vertigo. People have to move away when they are put up." He did not believe the sound from turbines was an issue, though some people did. He did say that when Beech Ridge performed audio studies to determine ambient sound, they left the recorders up all week, picking up wind and rain noise, and left them at noisy locations such as by a creek. They then said the ambient noise was louder than the turbines, that the natural sounds outweighed the sounds of the turbines. Stroud said that on a quiet night in the valley, with a breeze up on the ridge, the turbines could keep you awake.

Beech Ridge also had promoted the idea that there would be two hundred jobs required during site construction and had thus gained strong support

Endangered

from West Virginia unions. In his walks up on the ridge during construction, Stroud saw that, yes, local labor was being used for the clearing, for grading of sites and roads, and for laying the concrete. There could have been twenty to thirty people, he thought, maybe as many as fifty, but most of the license plates on the workers' vehicles were from Indiana, Texas, Utah, and so on. At church, Stroud met a woman whose husband, who was from Chicago, was working there but didn't want to admit what he did. The technical people were from out of town; only construction was local. With regard to Dave Groberg, who led the Invenergy effort to develop Beech Ridge, Stroud said, "I have nothing against him. In fact, in some of his testimony he was honest to the point that it was not to his company's advantage." Stroud concluded by saying, "I am not politically correct. I'm interested in energy but more interested in preserving the environment." He had become cynical about politics, he said, and now believed that wind energy is a "political fad."

What, then, is the future of wind energy?

Chapter 10

A View of the Mountain

ENERGY FUTURES

Wind energy in the United States had a relapse in 2010 and 2011 but bounced back in 2012 with a record 13,000 megawatts in added capacity (see Table 10.1). Eight thousand megawatts of that total were completed in the fourth quarter as developers raced to complete projects in 2012 to gain subsidies before they expired. Congress's one-year extension of the production tax credit through 2013 was helpful, but there were few wind projects in process at the time, so maintaining the 2012 rate of growth in the near future seemed unlikely. Despite lower costs, wind energy continues to be dependent on subsidies.

In 2012 wind provided 3.3 percent of the electricity generated in the United States, up from 2.9 percent a year earlier, according to the Energy Information Administration. In comparison, natural gas jumped from 20 percent of electricity generated to 30 percent during the same one-year period.

The United States' wind growth had peaked in 2009 at the low point of the great recession, when the entire economy was in trouble. But a large number of projects already underway were completed that year. Increased subsidies, tax advantages, and incentives to encourage the immediate startup of new projects to increase jobs did not get the anticipated results, at least not in 2010. Other factors overwhelmed wind. Obtaining financing was difficult despite loan guarantees. The demand for electricity took a dip due to the recession. Natural gas, with its increased reserves, low, stable cost, and lower emissions than coal, provided strong competition. And finally, the country still had no energy policy, no renewable energy standard that would further wind development.

In 2011 the federal government spent $16 billion in subsidies for the development of renewable energy, according to a Congressional Budget Office report. (In comparison, $2.5 billion was spent on tax breaks for the fossil-fuel industry.)[1] The cost of extending the production tax credit would be $3.3 billion a year.[2] (Other sources indicated $1 billion per year.) Some opponents took the position that the heavily subsidized wind energy industry represented only a small part of the country's total power generation and therefore these subsidies provided an "insufficient return" to taxpayers. But scientist Ryan Wiser of the Lawrence Berkeley National Laboratory said that without the tax credit, "the wind business falls off a cliff."[3]

Of importance to the economy, as touted by the Obama administration, were wind energy's eighty-five thousand jobs. The American Wind Energy Association stated that extending the tax credit would add fifty-four thousand jobs through 2016 and canceling the credit would result in the loss of thirty-seven thousand jobs.[4] (Putting these numbers in some perspective, the recent natural gas boom had resulted in six hundred thousand jobs.)

Politically, the future did not look bright for wind. Growth was directly related to subsidies, and the industry was aware that production tax credits could end and stimulus funds had deadlines. With the Republican majority in the House of Representatives, the major theme in Washington swung from

Table 10.1. U.S. Wind Energy Annual Growth by Year
(Installed Capacity in Megawatts)

Year	Annual	Cumulative	Percentage
2005	6,371	8,998	227
2006	2,454	11,452	27
2007	5,252	16,704	46
2008	8,362	25,066	50
2009	10,002	35,068	40
2010	5,214	40,282	15
2011	6,647	46,929	16
2012	13,078	60,007	28

Source: American Wind Energy Association, Fourth Quarter 2012 Market Report, January 20, 2013.

stimulus to debt reduction. Government spending, jobs, health care, and tax matters all had far more priority than energy policy or climate change. Would the production tax credits be renewed? As in the past, it depended on the capriciousness of politics.

With wind's costs becoming more in parity with other energy sources, the sensible approach, as endorsed by Republican senator Tom Coburn, was to extend the production tax credit for a five-year period but decrease it 20 percent annually. This way wind energy development could be planned, yet the developers would know that in five years the industry must stand on its own. Unfortunately, Coburn of Oklahoma, a state that supports wind energy, ended up voting against such a bill when it became entangled in provisions he opposed.[5]

Meanwhile, a Brooking Institute survey in early 2012 indicated that 62 percent of Americans believed in climate change. In 2008, shortly after the UN report on global warming, the number had been 75 percent. The low had been 50 percent in 2010, but that figure was on the rise due to "people . . . making connections in what they see in terms of weather."[6]

Table 10.2. U.S. Electricity Generation Mix Projections				
Source	Actual 2009 (%)	DE Projected 2030 (%)	EIA Projected 2035 (%)	DB Projected 2030 (%)
Coal	45	51	43	22
Natural Gas	23	7[a]	25	35
Nuclear	20	16	17	23
Wind/Solar	2	20	5	14
Renewable	8	6	9	6
Oil	2	—	1	—
Total	100	100	100	100

[a] New natural gas reserves not considered.

Notes: DE = Department of Energy; EIA = Energy Information Administration; DB = Deutsche Bank.

Sources: Department of Energy, Wind Power in America's Future; Energy Information Administration, "Annual Energy Outlook 2011"; Fulton and Mellquist, "Natural Gas and Renewables: A Secure Low Carbon Future Energy Plan for the United States." Deutsche Bank Climate Advisers, November 2010."

135

Ignoring politics and looking ahead, three energy forecasts predicted what the energy mix might be and what wind energy's role could be twenty years from now (see Table 10.2). The first projection, from the Department of Energy's *Wind Power in America's Future*, prepared in 2007, is a scenario of how wind can reach 20 percent of generation.[7] The report assumes a 39 percent increase in electricity consumption from 2005 to 2030, no specific government support (such as the production tax credit), no changes in wind technology, and a slight reduction in cost. Installed capacity reaches 300,000 megawatts, a level requiring space for wind farms equivalent to the whole state of West Virginia. Offshore capacity, situated along the East Coast, is 18 percent of the total. Wind's capacity factor exceeds 40 percent onshore and 50 percent offshore. Fossil-fuel and nuclear generation decrease as renewables increase. Since the report was done in 2007, however, it is somewhat out of date in not recognizing either the recent expansion in natural gas reserves or the continuing decrease in power demand resulting from the great recession of 2009. The report does recognize the lack of sufficient transmission lines as a major challenge.

A 2010 Energy Information Administration forecast projects the power demand in the United States to grow 25 percent by 2035.[8] The EIA, a politically independent government agency, appears to view energy's future far more realistically than advocates of renewables, including politicians who promote wind and solar power as the salvation for global warming. Coal is projected to decrease only slightly in percentage, but with the growth in demand this translates into more tons of coal used. The price of coal is advantageous, and the reserves remain extensive. Natural gas usage grows less than might be expected considering its recent increase in reserves, its low cost, and its lower emissions. Nuclear power decreases. (All three reports were prepared prior to the Fukushima disaster.) Renewables will have a significant jump, from 8 to 14 percent. That 14 percent breaks down into roughly 7 percent hydro, and though wind is projected to double in that period, it represents only 4 to 5 percent of the mix, far less than the 20 percent scenario. (The projection is based on production tax credits not being renewed after 2012.) Biomass and geothermal remain minor factors. Solar power, with its decreasing but still high cost and low capacity factor, grows sevenfold but still is less than 0.5 percent. This Energy Information Administration forecast would seem to indicate little if any reduction in emissions in the future.

A third forecast, to 2030, was issued by Deutsche Bank Climate Change Advisors and prepared in 2010.[9] It differs from the other two projections in that it sees only a minimal growth in electricity demand over the years and dramatically takes into account the new natural gas situation. The low

A View of the Mountain

growth in demand considers not only the dip related to the great recession but also the anticipation of reductions through conservation and energy efficiency. Natural gas forges ahead of coal as fossil fuels in total decrease relative to 2009. Nuclear power has a small increase. Wind reaches 14 percent (solar is included in the number but is an insignificant percentage). The remaining renewables, primarily hydro, have a slight decrease.

Thus the three reports show wind ranging from 20 percent to 5 percent of the energy mix in the future. The 20 percent scenario requires, as mentioned, 300,000 megawatts of installed capacity. Deutsche Bank's 14 percent requires 219,000 megawatts, a number that seems more feasible from 2011's base of 47,000 megawatts, thus requiring an average of 9,000 megawatts added each year. The United States added that much in 2009. China added slightly less than 20,000 megawatts of capacity in both 2010 and 2011. But the United States is not China with its authoritarian ways. This country's government must reach consensus on an energy policy and subsidize wind energy and the upgrade and expansion of the transmission system.

The Energy Information Administration forecasts that worldwide electricity demand will increase 87 percent between 2007 to 2035, largely driven by the needs of underdeveloped countries. Coal will barely increase as a percentage of the mix (see Table 10.3). But with the growth in demand, more tons of coal will be burned, emitting more carbon dioxide unless carbon capture develops rapidly. Renewables will be 23 percent of the mix, but that will be mostly hydro. Wind could be 6 to 7 percent. Nuclear power and natural gas will play a smaller role than in the United States. Again, it is difficult to see how energy emissions are going to be reduced if this projection comes true.

The Deutsche report does show how its energy mix in 2030 would reduce electric power emissions in the United States by 44 percent, a reduction in electrical power carbon emissions from 2.4 billion to 1.3 billion metric tons. The projection, as mentioned previously, is based on very little increase in demand. The expanded use of natural gas and the growth in wind/solar power contribute almost equally to three-quarters of the reduction. A switch from coal to nuclear power is a lesser factor, while coal efficiency makes a minor contribution. The 44 percent reduction in power emissions by 2030 fits somewhere between the Obama administration's (and the Democratic House of Representatives') overall economy reduction in greenhouse gas emissions of 17 percent by 2020 and 83 percent in 2050 from 2005 levels.

While a mix of energy sources tilted away from coal is crucial to meeting emission reductions, technology provides some promise of improvement in the long term. Coal, at least to some degree, is here to stay, but carbon

A View of the Mountain

Table 10.3. World Electricity Generation Mix Projections by Fuel

Source	Actual, 2007 (%)	Projected, 2035 (%)
Coal	42	43
Natural gas	21	19
Nuclear	14	13
Renewable	18	23
Oil	5	2
Total	100	100

Source: Energy Information Administration, "International Energy Statistics," 2010.

capture must be developed to reduce emissions. Wind and solar energy need storage capability, the ultimate battery, smart grids to damp fluctuations, and a national transmission system to move power where it is needed. Nuclear power needs to move ahead with a decision on waste storage and standardized plant designs that reduce cost and have shorter approval and construction times. Natural gas must assure that the Marcellus Shale fracturing process doesn't damage the water supply and the local environment. All this requires properly allocated research incentives.

And what seems lost is conservation. Home energy surveys, computerized household controls, more efficient bulbs and appliances, and better insulated homes need to be promoted, as they were in the 1970s, to encourage the use of less power. But the American people and their leaders don't have the will to any significant degree to face these matters, much less to make sacrifices.

More emphasis on talking about clean energy as opposed to just renewable energy is helpful. President Obama mentioned 80 percent clean energy in the future in his 2011 state of the union address. That number had to include not just renewables but nuclear and even fossil fuels, if carbon capture becomes a practical solution.

Because of politics, there is no national energy policy. There is no middle ground politically. There is no system study to balance what is most efficient. Subsidies and research funds are willy-nilly, determined by lobbyists influencing Congress. As Department of Energy secretary, Stephen Chu said engineers need to make these decisions, not politicians, but it doesn't work that way in this country. China has a long-term energy policy. They propose a mix of coal, nuclear, hydro, wind, and solar energy while at the same time subsidizing associated industries, all to cohesively meet the power demands of their growing economy. With politics the way they are in this country, however, such an energy plan seems out of reach. Whether or not one believes global warming is the threat it is portrayed as being, goals without plans and priorities to reduce emissions are little more than wishes.

THE IRONY

Far from a chart or table that projects the world's energy future, the country store in Frankfort, West Virginia, a few miles north of Lewisburg, has an interesting view. Several of the Beech Ridge turbines are easily seen several miles away, but there is something else far more noticeable on the heavily forested ridgeline. It is like a blight that has spilled over the ridge—an area that has been stripped of trees. The woman who ran the store didn't know what it was. Did it have something to do with the wind turbines? Was it a timbering operation or a quarry? A coal operation?

In another part of West Virginia, a recent wind-versus-coal situation occurred. It was depicted in the 2011 documentary *The Last Mountain,* which played to a full house at the old Lewis Theater in downtown Lewisburg. Coal River Mountain Watch, a group formed to protect one of the last mountains left in southern West Virginia from mountaintop removal, sponsored the movie. The group, as an alternative to mountaintop removal (taking the coal and destroying the surface), had financed a study and proposed that a 164-turbine, 328-megawatt wind farm be placed on the property. A great idea. Celebrities such as movie star Daryl Hannah and NASA climate scientist James Hansen supported the plan. The advantages were many: the mountain's contour would be saved, forests would be protected, there would be no fill dumped into the valleys or streams, local property tax income would increase significantly (coal companies paid little in local taxes), and employment, though lower than the employment coal would bring, would extend beyond the life of the mine. However, it came down to a matter of economics. The coal royalties to the property owner would far surpass the income from a wind farm. Without the owner's agreement, nothing could

A View of the Mountain

John Stroud at his farm in the Williamsburg Valley talks about his experience as co-chairman of Mountain Communities for Responsible Energy. (Photograph by the author.)

be done. The coal company began mining on a piece of the property, and the matter died.

Back in the Williamsburg Valley on the approach to John Stroud's farm, much closer to the ridge, it is obvious the apparition seen from Frankfort is a surface mine. Ledges and cuts can be seen. Through binoculars, the area looks like a moonscape. Is it mountaintop removal? The area to the west of the ridge has been mined, even strip mined, but no one has ever talked about mountaintop removal in this part of Greenbrier County, particularly overlooking the Williamsburg Valley.

Stroud and his neighbors, who had fought the wind turbines so diligently, did not know about the land that was to be stripped.[10] Their first notice came when large explosions from the excavation rattled Stroud's windows and shook his house. He had experienced an earthquake in California, and that's what the explosion felt like, he recalled. There had been no blasting notification. He and his neighbors "raised hell" and called the West Virginia Department of Energy. The intensity of the explosions subsided, but very little information was forthcoming about the mine. How much of the ridges would the mine scar? The mine was located on Plum Creek Timber Company land, hardly more than a mile from the point on the ridge above Stroud's house where Beech Ridge had originally planned to place a string of turbines before the federal judge intervened.

Plum Creek Timber, with a regional office in Lewisburg, is the largest landowner in the United States, holding seven million acres total, with over one hundred thousand of those acres in West Virginia. The company owned the surface above Williamsburg and had sold the mineral rights in that area to National Resource Partners in 2005 for $3.6 million.[11] National Resource Partners was a New York Stock Exchange master limited partnership, formed in 2002 with the purpose of acquiring producing mineral properties

A strip mine appears on the ridge above the Williamsburg Valley near the Beech Ridge wind farm site. (Photograph by the author.)

141

Beech Ridge turbines overlook a farm in Greenbrier County. (Photograph by the author.)

across the country. They had bought significant coal properties across West Virginia and throughout Appalachia.

Stroud said that the part of the ridge that had been mined had been seeded, adding some greenness to the bareness, but with the dry summer weather it had turned brown. Hardwood trees were to be planted, but John questioned whether they would survive since they were placed in a residue of loose gravel and rocks. With the mine so close to where the turbines would have gone and with the ridges all containing coal, if the mine moved north to MeadWestvaco property, he could have mountaintop removal directly above his farm. "I may have shot myself in the foot [by getting rid of the turbines]," he said.

Thus the irony. To protect the viewshed of the Williamsburg Valley, Mountain Communities has spent thousands of dollars over the course of more than five years presenting their case to the PSC, appealing to the state supreme court, and finally going before the federal court. Then a mountaintop-

A View of the Mountain

removal coal mine of more than a thousand acres appears on the ridgeline without warning, doing far more damage to the viewshed and the mountain ridges than the wind turbines, now looking like mere sticks in comparison, would ever do. Coal dominates.

And on a Florida beachfront, a nuclear power plant quietly provides power to a million homes. There is no room for wind.

Wind is the cover child, the leader of the band, in the attempt to prevent global warming, and it will have a role. But to face reality, fossil fuels, their carbon dioxide hopefully suppressed, and nuclear power, its risks more readily controlled and accepted, will prevail as the base load energy sources into the foreseeable future.

Epilogue

Five years after the Department of Energy's *Wind Power in America's Future,* another study determined that in 2050 the United States could receive 80 percent of its electrical generation from renewable resources. This is the conclusion of the 2012 Renewable Electricity Futures Study conducted by the National Renewable Energy Laboratory (NREL).[1] The study had looked at increasing renewables in 20 percent increments and determined that 80 percent renewables could be achieved. That number breaks down roughly into 40 percent from wind, 10 percent from solar, and 30 percent from other renewables. Coal, natural gas, and nuclear power provide the remaining 20 percent. And with 80 percent renewables, carbon emissions will be reduced 80 percent, according to the study.

This NREL conclusion is based on a variety of possible energy sources. There are more than ample renewable resources scattered across the United States, though the mix may vary by region—solar leads in the Southwest, hydro in the Northwest, wind in the Midwest, offshore wind in the East, and solar and biomass in the Southeast. With the renewables so scattered and diverse, a flexible grid and an expanded transmission system can handle the variability in generation and the fluctuations in demand on a national scale on an hour-to-hour basis. The technology is available now, and no potential technology advances were considered in the conclusion. In one scenario, it is assumed that savings from conservation and increased demand will balance each other, resulting in no growth in electricity demand over the next forty years. The cost impact is within a range that normally could be handled. Existing state and federal policies, including production tax credits, will continue until they expire (generally in 2017 or earlier). Land will be available through normal siting and permitting conditions. While wind and solar, with limited predictability, are not dispatchable on demand, fossil-fuel plants within the 20 percent non-renewables category provide the needed dispatchability and have sufficient generation capacity to provide "spinning" reserves. Storage capability is also a consideration, though the primary source for this seems vague beyond hydro and solar heating of fluids for overnight, short-term use. Government policy initiatives such as carbon cap and trade, carbon tax, or renewable portfolio standards (as already implemented by some states) are not considered in the study, the primary focus of which is technical issues.

As the study states, the analysis presented is far from complete: "Additional research would be required to more fully assess these challenges and opportunities, the market and regulatory changes that may be needed to facilitate such a transformation of the electricity system, as well as the costs and benefits of pursuing high levels of renewable electricity generation relative to non-renewable electricity generation options."[2]

The installed capacity necessary to reach the 80 percent renewable level requires roughly 500,000 megawatts for wind and 150,000 megawatts for solar energy.[3] Wind capacity grew from 2,000 megawatts in 2000 to 60,000 in 2012. Solar capacity reached 3,000 in 2011. Wind will have to expand at an average rate of almost 12,000 megawatts per year over forty years to reach 500,000 megawatts. Its growth in the United States has varied with the availability of federal subsidies, peaking at 9,000 megawatts in 2009, dropping to 5,000 megawatts in 2010 and 6,000 megawatts in 2011, and hitting an all time high of 13,000 in 2012. A decrease from that level in 2013 is anticipated. Meanwhile, China expanded its wind capacity by 18,000 megawatts in 2011. So reaching the necessary wind capacity in 2050 is, perhaps, feasible.

The NREL study says that the expansion of renewable facilities, including transmission lines, "totaled less than 3% of the land area of the contiguous United States." To put that number in some kind of perspective, the wind farms, solar plants, and transmission lines necessary to meet the 80 percent figure would require a land area equivalent to the combined square miles of the states of West Virginia, Virginia, and North Carolina. If "not in my back yard" is a concern today and travelers crossing the Midwest even in 2012 are amazed at the number of wind turbines they see, think what it could be like forty years from now. On the other hand, if it becomes more apparent that global warming is dramatically affecting the country, then the NIMBY issue may become a moot point.

The key word throughout the NREL study is "challenge." Predicting a condition forty years in advance is fanciful. The NREL study presents from an authoritative source the convincing picture that in a perfect world the technology is there to accomplish a goal of 80 percent renewable energy.

Does looking at forty years of past history provide clues to the future of energy? Since 1970, we have been involved in four wars—Vietnam, the Iraq wars, and Afghanistan—all absorbing national funds and causing debt. We have had five Republican presidents and three Democrats, providing various approaches to energy. Excluding the wars, this country has completed only one major national endeavor—putting a man on the moon. Yet through innovation (both government and private) and free enterprise, the personal computer, the cell phone, and the Internet have been developed during this period.

From an energy standpoint, we had in the 1970s the peak in nuclear power plant construction, the Arab oil embargo, and the Three Mile Island nuclear accident. In the 1980s came the California wind energy boom and bust, the Chernobyl nuclear disaster, and the beginning of warnings about carbon emissions and climate change. In the 1990s Congress passed the Energy Policy Act, which deregulated the electric power industry and encouraged the use of alternative energy. This sparked the development of wind farms by independent investors who with subsidies could sell the electricity to large utilities on a profitable basis. The Kyoto Protocol was initiated, encouraging countries to pledge to meet emission reduction goals. Though ignored by the United States, Kyoto resulted in the development of wind energy, particularly in Europe, where a carbon cap and trade market also was established. In 2007 a UN scientific panel concluded that man was probably the cause of global warming. Congressional Democrats and President Obama supported this theory; the Republicans did not. Meanwhile, wind energy grew at a rapid rate both in Europe and the United States. The great recession of 2009 led to a stimulus package that encouraged further development of renewables. A proposed cap and trade market failed in Congress. Wind energy growth peaked in 2009 but then leveled off as the stimulus incentives expired. The cost of solar cells decreased rapidly, giving solar energy a boost, though it still remained a minor resource. Almost out of nowhere, deep reserves of natural gas in the United States were found and the technology was available to access this suddenly plentiful resource. Natural gas, with half the carbon emissions of coal plus lower cost, began to have an impact on both coal and wind usage. And elsewhere, China became the world leader in wind and solar development, moving ahead of Europe with its economic problems and the United States with its political stalemate.

So what will it take for 80 percent renewables in 2050? The technology may already be there, but achieving such a goal will require leadership, political will, a robust economy, and a national effort like creation of the 1930s

Rural Electrification Administration, the 1940s atomic bomb development, the 1950s-initiated interstate highway system, and the 1960s space program. Perhaps 80 percent renewable energy would be most like the interstate highway program, which required the federal government to lead, the states to implement, and the people to support. But adding nationwide more renewable resources (particularly wind and solar), plus transmission lines and smart, flexible grids is far more complex. It requires a transformation of our energy infrastructure through the coordination of federal, state, industrial, and local interests. A climate change disaster or the stronger threat of such disaster might be what it takes to bring the country together to achieve this challenge—as World War II did for the gigantic industrial effort that helped win that war.

On September 25, 2012, Beech Ridge Energy announced that the company would add thirty-three turbines to the west of the sixty-seven turbines already in place in Greenbrier County. Apparently the environmental conditions imposed by the federal judge concerning bats were on their way to being resolved. The additional terminals would be in place by the end of 2013.

Notes

1. FROM THE BEACH TO THE MOUNTAINS

1. Henrietta McBee, Dec. 7, 2007. Unless otherwise noted, all activities and statements were witnessed by the author.

2. The author toured the Tucker County area July 17–18, 2007, talking with various residents.

3. "County Windmills Generate Attention," *Parsons Advocate*, Apr. 13, 2005.

4. Dennis Moloney, interview with the author, Aug. 12, 2009. The author accompanied Moloney on his inspection of the construction site.

5. *STAAD.pro 2007 Design of Wind Turbine Foundations*, brochure, AES Corporation, Bentley RAM/STAAD Solution Center, Nov. 2, 2007.

6. *1.5MW Wind Turbine*, brochure, GE Energy, 2009.

7. "Betz Limit," Renewable Energy UK, http://www.reuk.co.uk/Betz-Limit.htm/.

8. "Fundamentals of Wind Energy," presented at the American Wind Energy Association fall symposium, Orlando, Fla., Nov. 18, 2009.

2. A THOUSAND YEARS

1. Robert W. Righter, *Wind Energy in America: A History* (Norman: Univ. of Oklahoma Press, 1990), 9.

2. Ibid., 21.

3. Paul Gipe, *Wind Power: Renewable Energy for Home, Farm, and Business* (White River Junction, Vt.: Chelsea Green, 2004), 83.

4. Peter Asmus, *Reaping the Wind: How Mechanical Wizards, Visionaries, and Profiteers Helped Shape Our Energy Future* (Washington D.C.: Island Press, 2001), 40.

5. Righter, *Wind Energy in America*, 128.

6. Gipe, *Wind Power*, 352.

7. Righter, *Wind Energy in America*, 173.

8. Asmus, *Reaping the Wind*, 126.

9. Ibid., 94.

10. Ibid., 95.

11. Ibid., 117.

12. Righter, *Wind Energy in America*, 171.

13. Asmus, *Reaping the Wind*, 85.

14. Ibid., 216.

3. Europe: Reality or Tilting at Windmills

1. "Our Planet (Denmark/Wind Power)," *NBC Evening News*, Nov. 5, 2007.

2. "Energy Efficiency from the Wind," *U.S. News & World Report*, Mar. 18, 2007.

3. Dr. V. C. Mason, "Wind Power in Denmark," last modified Mar. 2007, *Country Guardian*, http://www.countryguardian.net/.

4. Hugh Sharman, Henrik Meyer, and Martin Agerup, *Wind Energy: The Case of Denmark* (Copenhagen: Center for Politiske Studier, Sept. 2009).

5. "The Danish Wind Case: Fast Facts," report, Energinet.dk, Oct. 2008, https://selvbetjening.preprod.energinet.dk/NR/rdonlyres/34B90E46-A9DA-4F5C-B5D1-72ADF79C1991/0/TheDanishWindCaseFastFactsUKversion.pdf/.

6. "Wind Energy in Germany 2009," German Wind Energy Association, Copy in author's possession.

7. "An Ill Wind Blows across Germany," *Age*, Oct. 12, 2007.

8. E.ON Netz, "Wind 2005" and 2006 supplement. http://www.nerc.com/docs/pc/ivgtf/EON_Netz_Windreport2005_eng.pdf, Aug. 31, 2006.

9. "Germany's Wind Farms Challenged," *BBC World Service*, May 28, 2006.

10. "Will Nuke Phase-out Make Offshore Farms Attractive?" *Spiegel Online*, Mar. 23, 2011, www.http://www.spiegel.de/.

11. European Wind Energy Association, "The European Offshore Wind Industry Key 2011 Trends and Statistics," European Wind Energy Association, Jan. 2012, updated July 2012, http://www.ewea.org/fileadmin/files/library/publications/statistics/EWEA_stats_offshore_2011_02.pdf/.

12. "Germany OKs Huge Offshore Wind Farms," *Business Week*, Sept. 21, 2009.

13. "Will Nuke Phase-out Make Offshore Farms Attractive?"

150

14. "Spain's Wind Power Industry on a Roll," Renewable Energy Access, Aug. 24, 2005.

15. "Economy of Spain on the Edge," *New York Times*, Mar. 2010.

16. "AEE Announces High Risk of Job Loses Due to Wind Sector Paralysis," Eolic Energy, Dec. 23, 2009, http://eolicenergynews.org/, Copy in author's possession.

17. "People in the Countryside Told to Accept 'Many Thousands' of Wind Turbines as Part of New Energy Strategy," *London Telegraph*, July 16, 2009.

18. "Wind Power Plants 'Flawed,' Say Critics," *London Telegraph*, July 18, 2009.

19. "Wind Farms Produced 'Practically No Electricity' during Britain's Cold Snap," *London Telegraph*, Jan. 11, 2010.

20. "Renewables Obligation," Ofgem, last modified Apr. 16, 2010, http//:ofgem. gov.uk/., Copy in author's possession.

21. "The London Array: The World's Largest Offshore Farm," *London Telegraph*, July 28, 2012.

22. "European Offshore Wind Industry Key 2011 Trends and Statistics."

23. "Europe Is Missing Carbon Cut Targets," *London Telegraph*, June 15, 2007.

24. "Is Europe Really on Track to Meet Its Kyoto Goals?" *New Republic*, Nov. 11, 2009.

25. "Breathing Difficulties, a Market in Need of a Miracle," *Economist*, Mar. 3, 2012.

4. BATTLE BEGINS

1. "Wind Turbine Electricity coming to Greenbrier," *Register Herald* (Beckley, W.Va.), July 13, 2005.

2. Debbie Sizemore, interview with the author, Feb. 28, 2007. Lewisburg, West Virginia.

3. "Wind Factories Called Corporate Hoax," *Mountain Messenger* (Lewisburg, W.Va.), Oct. 1, 2005.

4. Alice Crowe, letter to the editor, Sizemore Scrapbook. Debbie Sizemore loaned the author her scrapbook of newspaper clippings on February 28, 2007.

5. Sizemore interview.

6. "Wind Energy Controversy Grows," *Mountain Messenger*, Oct. 8, 2005.

7. John Stroud, interview with the author, Oct. 19, 2006. Williamsburg, West Virginia.

8. "Wind Energy, Clean Energy or Corporate Boondoggle," PowerPoint presentation, Mountain Communities for Responsible Energy, obtained from John Stroud, Sept. 30, 2006.

9. "Wind Factories Called Corporate Hoax."

10. "Polsky Divorce in Chicago Could Strike a Record," *Chicago Tribune*, June 4, 2007.

11. "Invenergy Wind," Invenergy LLC, last modified June 20, 2006, http//:www.invenergyllc.com/.

12. Dave Groberg, telephone interview with the author, Oct. 1, 2007.

13. "Title 150 Legislative Rule Public Service Commission, Series 30 Rules Governing Siting Certificates for Exempt Wholesale Generators," West Virginia Legislature, 2006, reviewed, Feb. 20, 2007. http://www.psc.state.wv.us/rules.htm. Copy in author's possession.

14. Public Service Commission of West Virginia, "Beech Ridge Energy LLC, Application for a Siting Certificate, Case 05-1590-E-CS," Public Service Commission of West Virginia, Charleston, Nov. 1, 2005.

15. Groberg interview.

16. "Citizen Files 'Petition to Intervene' in Wind Farm Project," *Mountain Messenger*, Dec. 5, 2005.

17. John Walkup, letter to the editor, *Mountain Messenger*, Apr. 22, 2006.

18. Eddie Fisher, letter to the editor, Sizemore Scrapbook.

19. Glenn McKinney letter to the editor, Sizemore Scrapbook.

20. "Greenbrier CVB Opposes Wind Farm," *West Virginia Daily*, Sizemore Scrapbook.

21. Gary Smith, letter to the editor, Sizemore Scrapbook.

22. "The Greenbrier Joins Wind Project Fight," *Charleston Gazette*, May 17, 2006.

23. "Bills Not Favorable to Proposed Wind Farm," *Register Herald*, Feb. 23, 2006.

24. "Concerning Proposed Wind Farm Project," *West Virginia Daily News*, Sizemore Scrapbook.

25. "Wind Energy Company Seeks Public Hearing," *Register Herald*, Dec. 19, 2005.

26. Beech Ridge–Public Service Commission hearing, Apr. 2006, Lewisburg, W.Va., attended by the author; "Residents Say Wind 'Factory' Would Spoil Life," *Charleston Gazette*, Apr. 26, 2006; "Hearing Held on Proposed Wind Farm for Greenbrier County," *Valley Ranger* (Lewisburg, W.Va.), Apr. 30, 2006.

27. "Voter Opinion Survey, Final Results 04.18.06," RMS Strategies, Charleston, W.Va., provided by Dave Groberg, Nov. 1, 2007.

5. Rough Road

1. Skip Deegans, telephone conversation with the author, Oct. 26, 2006.

2. Public Service Commission of West Virginia, "Siting Certificate for Beech Ridge Energy, Case NO. 05-1590-E-CS," Public Service Commission of West Virginia, Charleston, Aug. 28, 2006. Items and issues discussed during the hearings are based on the Public Service Commission order.

3. "Moratorium Move 'Shocks' Wind Farm Developers," *Register Herald,* June 11, 2006.

4. The author attended the Lewisburg City Council meeting.

5. "Midland Trail Scenic Highway Assoc. Receives $10,000 Gift," *Valley Ranger,* July 12, 2006.

6. "Siting Certificate for Beech Ridge Energy, Case NO. 05-1590-E-CS," 34.

7. "Petition for Reconsideration of the Commission's August 28, 2006 Order Granting Beech Ridge's Application," Mountain Communities for Responsible Energy, Williamsburg, W.Va., Sept. 18, 2006, 2-24, 26-40.

8. John Stroud, telephone interview with the author, Feb. 5, 2007.

9. Sizemore Scrapbook.

10. "CO2 Emissions in U.S. Drop to 20-year Low," Associated Press, *Charleston Gazette,* Aug. 17, 2012.

11. "Question of the Week," *Business Week,* Feb. 19, 2007.

12. "Climate Change: Carbon Dioxide Levels in World's Air Reach 'Troubling Milestone,'" *Huffington Post,* May 31, 2012, http://www.huffingtonpost.com/2012/05/31/climate-change-carbon-dioxide-troubling-milestone_n_1558561.html/.

13. "Game Over for the Climate," James Hansen, editorial, *New York Times,* May 10, 2012.

14. "Adapting to the Impacts of Climate Change," Report in Brief, National Academy of Sciences, May 2010, http://dels.nas.edu/resources/static-assets/materials-based-on-reports/reports-in-brief/Adapting_Report_Brief_final.pdf/.

15. "Wind Farm Foes, Proponents Pitch Arguments to the Supreme Court," *Register Herald,* Jan. 9, 2008.

16. Bradley W. Stephens, attorney, Mountain Communities for Responsible Energy, to Lee F. Feinberg, attorney, Beech Ridge Energy, Dec. 10, 2008.

17. Lee F. Fienberg, Beech Ridge Energy, to Sandra Squire, Executive Secretary, Public Service Commission, Dec. 12, 2008.

153

18. Public Service Commission of West Virginia, "Commission Order, Beech Ridge Energy LLC," Public Service Commission of West Virginia, Charleston, Feb. 13, 2009.

19. "State PSC Won't Consider Beech Ridge Wind Farm OK," *Register Herald*, Apr. 3, 2009.

6. WIND IN THE STATES

1. American Wind Energy Association, "AWEA U.S. Wind Industry Annual Market Report Year Ending 2011," Jan. 2012.

2. "Texas Oilman T. Boone Pickens Wants to Supplant Oil with Wind," *USA Today*, July 11, 2008.

3. "Texas Is More Hospitable than Mass. to Wind Farms," *Boston Globe*, Sept. 25, 2006.

4. CNET News, October 1, 2009, Copy in author's possession.

5. "Report: Iowa First in Wind Energy Percentage," *Ames Tribune*, Apr. 8, 2010.

6. "Texas Oilman T. Boone Pickens."

7. "Hartland Wind Farm Is a Work-in-Progress," *Kenmare News*, Nov. 11, 2011.

8. "Wind Power Generates a New Cash Crop in State," *Seattle Times*, Jun. 19, 2006.

9. "The Big Picture," California Wind Energy Association, last modified June 13, 2010, http://calwea/.

10. "Tax Dollars Blow Away in Wind Projects," *Oregonian*, Nov. 28, 2009.

11. Ibid.

12. "Wind Farm 'Mega-Project' Underway in Mojave Desert," *Los Angeles Times*, July 27, 2010.

13. New York State Energy Research and Development Authority, "Frequently Asked Questions," New York State Energy Research and Development Authority, last modified July 1, 2010, http://www.nyserda.ny.gov/About/Frequently-Asked-Questions.aspx/.

14. "Noble Environmental Power Overcomes Market Challenges, Receives Long-Term Capital from GE and Others for 3 NY Windparks," Noble Environmental Power, last modified Apr. 6, 2009, http://noblepower.com/.

15. "Company Working to Turn Fenner Wind Turbines on in July," WSYR-TV, Syracuse, N.Y., last modified June 22, 2011, http://wsyr.com/.

16. "How Gov. Crist Became Gov. Climate," *St. Petersburg Times*, July 21, 2007.

17. "Rates May Soar If Green Electric Bills Are Passed," *Tampa Tribune*, Apr. 8, 2008.

18. "Charlie Crist Superstar," *Grist*, Jan. 14, 2008.

19. "Turbine Plan Switch 'Didn't Solve Anything,'" *Scripps Treasure Coast Newspapers*, Jan. 18, 2008.

20. Wendy Williams and Robert Whitcomb, *Cape Wind* (New York: Public Affairs, 2007), 95.

21. "Gov't Ok's 1st US Offshore Wind Farm off Mass," Yahoo! News (Associated Press), last modified Apr. 28, 2010, http://news.yahoo.com/.

22. "Cape Wind to Sell 50% of Offshore Output to National Grid,' *Wall Street Journal*, May 7, 2010; U.S. Energy Information Administration, "Average Retail Price of Electricity to Ultimate Customers by End-Use Sector by State," U.S. Energy Information Administration, last modified June 16, 2010, http://eia.doe.gov/.

23. "FAA Rules Cape Wind Will Not Affect Air Traffic," *Boston Globe*, Aug. 16, 2012.

7. Alternative Sources

1. "Is America Ready to Quit Coal?" *New York Times*, Feb. 15, 2009.

2. "NASA Scientist Calls for Phase-out of Coal Use by 2030," *Charleston Gazette*, June 24, 2008.

3. "W.Va. Charts Coal's Future?" *Charleston Gazette*, Oct. 31, 2009.

4. "Tracking New-Coal Fired Power Plants," National Energy Technology Laboratory, last modified Jan. 8, 2010, http://netl.gov/.

5. Secretary Stephen Chu, Department of Energy, "The Roadmap for Carbon Sequestration," PowerPoint presentation, Carbon Capture and Sequestration Forum, Univ. of Charleston, Charleston, W.Va., Sept. 8, 2010, provided by Senator's Jay Rockefeller's office, Sept. 13, 2010.

6. "FPL Highlights Strategy to Meet Florida's Future Electric Supply Needs," Florida Power and Light, Apr. 4, 2007; "Ten Year Power Plant Site Plan, 2011–2120," Executive Summary, 2011, Florida Power and Light, Copy in author's possession. https://www.frcc.com/Planning/Shared%20Documents/FRCC%20 Presentations%20and%20Utility%2010-Year%20Site%20Plans/2011/2011_TYSP_ FPL.pdf/.

7. "Utilities Turn from Coal to Gas, Raising Risk of Price Increase," *New York Times*, Feb. 5, 2008.

8. "A Burning Issue," *Economist*, Jan. 28, 2012.

9. "Nuclear Retreat to Add 30 Percent to Co2 Growth: IEA," Reuters, June 16, 2011.

10. "Chernobyl: The True Scale of the Accident," WHO/IAEA/UNDP joint news release, Sept. 5, 2005, World Health Organization, http://www.who.int/mediacentre/news/releases/2005/pr38/en/index.html/.

11. "When the Seam Clears, *Economist*, May 24, 2011.

12. Gwyneth Cravens, *Power to Save the World: The Truth about Nuclear Energy* (New York: Alfred A. Knopf, 2007), 2005.

13. Nuclear Energy Institute, *Used Nuclear Fuel: Handled with Care: How the Nuclear Energy Industry Manages Used Fuel on Site* (Washington, D.C.: Nuclear Energy Institute, Apr. 2005).

14. Jon Gertner, "Atomic Balm?" *New York Times Magazine*, July 16, 2006, 62, 64.

15. "Japan Utilities Emit Record CO2 After Fukushima Disaster," Bloomberg. com, Aug. 6, 2012, http://www.bloomberg.com/news/2012-08-09/japan-utilities-emit-record-co2-after-fukushima-disaster.html/.

16. Gertner, "Atomic Balm?" 56, 64.

17. "Facing Up to Nuclear Risk," *Bloomberg Businessweek*, Mar. 21, 2011.

18. Paul Gipe, "New Record for German Renewable Energy in 2010," Renewable Energy World, Mar. 25, 2011, http://www.renewableenergyworld.com/rea/news/article/2011/03/new-record-for-german-renewable-energy-in-2010/.

19. Alexander Neubacher and Catalina Schröder, "Germans Cough Up for Solar Subsidies," Spiegel Online, July 4, 2012, http://www.spiegel.de/international/germany/german-solar-subsidies-to-remain-high-with-consumers-paying-the-price-a-842595.html/.

20. "Asia to Overtake Europe as Global Solar Power Grows EPIA," Reuters, May 7, 2012.

21. "The Newest Hybrid Model," *New York Times*, Mar. 4, 2010.

22. "Solar Research Frequently Asked Questions," Florida Power and Light, last modified Sept. 28, 2010, http://fpl/com/. Copy in author's possession.

23. "Mojave Solar Park," California Photon, last modified July 10, 2010, http://californiaphoton.com/.

24. "APNewsbreak: Energy Dept. Announces $4.5 Billion in Loan Guarantees for Calif. Solar Projects," *Washington Post*, June 30, 2011.

25. "$1 Billion Solar-Thermal Plant near Gila Bend to Supply APS Customers," *Arizona Republic*, Feb. 21, 2008.

26. "Ties to Obama Aided in Access for Big Utility," *New York Times*, Aug. 22, 2012.

8. Growth Issues

1. U.S. Department of Energy, *Wind Power in America's Future: 20% Wind Energy by 2030* (New York: Dover Publications, 2010), 105, 106, 116.

2. "Town Councilor Regrets High Sheldon Wind Farm," RiverCityMalone.com, last modified July 10, 2010, http://rivercitymalone.com/.

3. "Maple Ridge Wind Farm Has Been a Disaster," *Watertown Daily Times,* Mar. 26, 2009.

4. "Lawsuits, Altoona Accident End Wind Farm in Beekmantown," *Press Republican* (Plattsburgh, N.Y.), Mar. 16, 2010.

5. "Stamford Gets Wind Opinions," *Daily Star* (Oneonta, N.Y.), Dec. 25, 2006.

6. Remarks by Mike Triolo, Dave Groberg to author, e-mail, Nov. 1, 2007.

7. "On an Upstate Wind Turbine Project, Opinions as Varied as the Weather, *New York Times,* Oct. 28, 2007.

8. Department of Energy, *Wind Power in America's Future,* 95, 96.

9. "Wind Power," Xcel Energy, http://xcelenergy.com/.

10. "Minnesota Wind Integration Study," report, Minnesota Public Utilities Commission, St. Paul, Nov. 30, 2006. http://www.uwig.org/windrpt_vol%201.pdf. Matt Schuerger, technical adviser to the Minnesota Public Utility Commission, provided some understanding and clarifications of the study to the author in a telephone conversation on September 24, 2007.

11. "West Texas Breezes Push Wind to Set a Record on the Grid," *Fort Worth Star-Telegram,* Mar. 1, 2010.

12. "Getting to $787 Billion," *Wall Street Journal,* Feb. 17, 2009.

13. "Adding Something for Everyone, House Leaders Won Climate Bill," *New York Times,* July 1, 2010.

14. "Republican Delegation Seizes on Energy Bill as 'National Tax,'" TheHill.com, modified Apr. 26, 2009, http://thehill.com/.

15. "As Climate Summit Nears, Skeptics Gain Traction," *Time,* Dec. 2, 2009.

16. "Tax Dollars Blow Away in Wind Projects," *Oregonian,* November 28, 2009.

9. Endangered

1. "Groups Seek Federal Ruling on WV Wind Farm," *Chicago Tribune,* June 11, 2009.

2. "MCRE File Lawsuit against Beech Ridge to Halt Wind Turbine Construction," *Mountain Messenger,* July 18, 2009.

3. "Bat-gate: Cover-up at the Beech Ridge Wind Facility," Wind Action, Oct. 9, 2009, http://www.windaction.org/faqs/23513/.

4. "Tiny Bat Pits Green against Green," *Washington Post*, Oct. 22, 2009.

5. Dave Groberg to author, e-mail, Oct. 27, 2009; John Stroud, telephone interview with the author, Oct. 28, 2009.

6. *Animal Welfare Institute, et al., v. Beech Ridge Energy LLC, et al.*, Order, Case No. RWT 09cv1519, United States District Court for the District of Maryland, Dec. 8, 2009.

7. *Animal Welfare Institute, et al., v. Beech Ridge Energy LLC, et al.*, Stipulation, Case No. 09-1519 (RWT), United States District Court for the District of Maryland, Jan. 19, 2010.

8. John Stroud, interview with the author, Sept. 22, 2010.

9. Stroud interview, Sept. 22, 2010.

10. A View of the Mountain

1. "The U.S. Subsidizes All Forms of Energy, including Fossil Fuels," *Charleston Gazette*, July 16, 2012.

2. "Amid a Political Calm, a Tax Break for the Wind Industry Advances," *New York Times*, Aug. 2, 2012.

3. "Tax Credit in Doubt, Wind Power Industry Is Withering," *New York Times*, Sept. 20, 2012.

4. "A Campaign Compromise on Wind Power," *Bloomberg Businessweek*, Aug. 23, 2012.

5. "Amid Political Calm."

6. "More Americans Now Believe in Global Warming," *Los Angeles Times*, Feb. 29, 2012.

7. Department of Energy, *Wind Power in America's Future*.

8. U.S. Energy Information Administration, "Annual Energy Outlook 2011," report, U.S. Energy Information Administration, Washington, D.C., Dec. 2010.

9. Mark Fulton and Nils Mellquist, "Natural Gas and Renewables: A Secure Low Carbon Future Energy Plan for the United States," Deutsche Bank Climate Change Advisors, Frankfurt, Germany, Nov. 2010.

10. Stroud interview, Sept. 22, 2010.

11. "Deed of Conveyance between Plum Creek Timberlands and WWPP LLC," Greenbrier County, W.Va., Feb. 25, 2005, Deed 496, p. 168.

1. National Renewable Energy Laboratory, "Renewable Electricity Futures Study," National Renewable Energy Laboratory, Golden, Colo., June 2012. The NREL is a part of the Department of Energy but is managed by two independent, nongovernment research organizations, MRI Global and Battelle, under a $1.1-billion, five-year contract.

2. Ibid., li.

3. Ibid., Figure ES-3, "Installed capacity and generation on 2050," xxx.

Selected Bibliography

PRIMARY SOURCES

American Wind Energy Association. "Summary of the American Recovery and Reinvestment Act (ARRA) of 2009." American Wind Energy Association, modified Dec. 14, 2009. http://awea.com/. Copy in author's possession.

———. "U.S. Wind Energy Projects." American Wind Energy Association, modified June 22, 2010. http://www.awea.org/. Copy in author's possession.

Animal Welfare Institute, et al., v. Beech Ridge Energy LLC, et al. Order. Case No. RWT 09cv1519. United States District Court for the District of Maryland, Dec. 8, 2009.

Animal Welfare Institute, et al., v. Beech Ridge Energy LLC, et al. Stipulation. Case No. 09-1519 (RWT). United States District Court for the District of Maryland, Jan. 19, 2010.

"Betz Limit." Renewable Energy UK. http://www.reuk.co.uk/Betz-Limit.htm/.

"Chernobyl: The True Scale of the Accident." WHO/IAEA/UNDP joint news release. Sept. 5, 2005. World Health Organization. http://www.who.int/mediacentre/news/releases/2005/pr38/en/index.html/.

Chu, Secretary Stephen. Department of Energy. "The Roadmap for Carbon Sequestration." PowerPoint presentation, Carbon Capture and Sequestration Forum, Univ. of Charleston, Charleston, W.Va., Sept. 8, 2010. Provided by Senator's Jay Rockefeller's office, Sept. 13, 2010.

"The Danish Wind Case: Fast Facts." Report. Oct. 2008. Energinet.dk. https://selvbetjening.preprod.energinet.dk/NR/rdonlyres/34B90E46-A9DA-4F5C-B5D1-72ADF79C1991/0/TheDanishWindCaseFastFactsUKversion.pdf/.

"Deed of Conveyance between Plum Creek Timberlands and WWPP LLC." Greenbrier County, West Virginia, Feb. 25, 2005. Deed 496, p. 168.

E.ON Netz. "Wind 2005." http://www.nerc.com/docs/pc/ivgtf/EON_Netz_Windreport2005_eng.pdf, Aug. 31, 2006.

European Wind Energy Association. "The European Offshore Wind Industry Key 2011 Trends and Statistics." Report. European Wind Energy Association, Jan. 2012, updated July 2012. http://www.ewea.org/fileadmin/files/library/ publications/statistics/EWEA_stats_offshore_2011_02.pdf/.

Fulton, Mark, and Nils Mellquist. "Natural Gas and Renewables: A Secure Low Carbon Future Energy Plan for the United States." Deutsche Bank Climate Change Advisors, Frankfurt, Germany, Nov. 2010.

"Fundamentals of Wind Energy." Presented at the American Wind Energy Association fall symposium, Orlando, Fla. Nov. 18, 2009.

GE Energy. *1.5MW Wind Turbine.* Brochure. GE Energy, 2009.

Global Wind Energy Council. "Global Installed Wind Power Capacity." Global Wind Energy Council. Copy in author's possession.

"Inventory of U.S. Greenhouse Gas Emissions and Sinks: 1990–2009." U.S. Environmental Protection Agency, Washington, D.C., Apr. 15, 2011. http://www. epa.gov/climatechange/Downloads/ghgemissions/US-GHG-Inventory-2011-Complete_Report.pdf/.

National Renewable Energy Laboratory. "Renewable Electricity Futures Study." Report. National Renewable Energy Laboratory, Golden, Colo., June 2012.

National Research Council. "Adapting to the Impacts of Climate Change, Report in Brief." National Academy of Sciences, May 2010. Copy in author's possession.

New York State Energy Research and Development Authority . "Frequently Asked Questions." Last modified July 1, 2010. New York State Energy Research and Development Authority. http://www.nyserda.ny.gov/About/Frequently-Asked-Questions.aspx /.

"Petition for Reconsideration of the Commission's August 28, 2006 Order Granting Beech Ridge's Application." Mountain Communities for Responsible Energy, Williamsburg, W.Va., Sept. 18, 2006.

Public Service Commission of West Virginia. "Beech Ridge Energy LLC, Application for a Siting Certificate, Case 05-1590-E-CS." Public Service Commission of West Virginia, Charleston, Nov. 1, 2005.

———. "Commission Order, Beech Ridge Energy LLC." Public Service Commission of West Virginia, Charleston, Feb. 13, 2009.

———. "Siting Certificate for Beech Ridge Energy, Case NO. 05-1590-E-CS." Public Service Commission of West Virginia, Charleston. Aug. 28, 2006.

"Renewable Energy Capacities by Country." Green Chip Stocks. Mar. 12, 2010. http://www.greenchipstocks.com/articles/renewable-energy-capaci-ties-by-country/765/.

"Renewable Energy—Wind, Cumulative Installed Wind Turbine Capacity." BTM Consult ApS. Last modified July 15, 2011, http://www.btm.dk/.

Sharman, Hugh, Henrik Meyer, and Martin Agerup. *Wind Energy: The Case of Denmark.* Copenhagen: Center for Politiske Studier, Sept. 2009.

"Solar Research Frequently Asked Questions." Florida Power and Light. Last modified Sept. 28, 2010. http://fpl/com/. Copy in author's possession.

"Ten Year Power Plant Site Plan, 2011–2020." Executive Summary. 2011. Florida Power and Light. https://www.frcc.com/Planning/Shared%20Documents/FRCC%20Presentations%20and%20Utility%2010-Year%20Site%20Plans/2011/2011_TYSP_FPL.pdf/.

"Title 150 Legislative Rule Public Service Commission, Series 30 Rules Governing Siting Certificates for Exempt Wholesale Generators." West Virginia Legislature, 2006, reviewed Feb. 20, 2007. Copy in author's possession.

Union of Concerned Scientists. "Public Utility Regulatory Policy Act (PURPA)." Last modified Nov. 8, 2007. Union of Concerned Scientists. http://www.ucsusa.org/clean_energy/smart-energy-solutions/strengthen-policy/public-utility-regulatory.html/.

U.S. Department of Energy. *Wind Power in America's Future: 20% Wind Energy by 2030.* New York: Dover Publications, 2010.

U.S. Energy Information Administration. "Annual Energy Outlook 2011." Report. U.S. Energy Information Administration, Washington, D.C., Dec. 2010.

———. "Average Retail Price of Electricity to Ultimate Customers by End—Use Sector by State." U.S. Energy Information Administration, Washington, D.C., June 16, 2010. http://eia.doe.gov/.

———. "Emissions of Greenhouse Gasses in the United States—2008." U.S. Energy Information Administration, Washington, D.C., Dec. 2009.

———. "International Energy Statistics." 2010. U.S. Energy Information Administration, Washington, D.C. http://www.eia.gov/cfapps/ipdbproject/IEDIndex3.cfm?tid=2&pid=2&aid=2/.

———. Office of Energy Analysis. "Updated Capital Cost Estimates for Electricity Generation Plants." U.S. Energy Information Administration, Nov. 2010. http://www.eia.gov/oiaf/beck_plantcosts/pdf/updatedplantcosts.pdf/.

"Voter Opinion Survey, Final Results 04.18.06." RMS Strategies, Charleston, W.Va. Provided by Dave Groberg, Nov. 1, 2007.

"Wind Energy, Clean Energy or Corporate Boondoggle." PowerPoint presentation. Mountain Communities for Responsible Energy. Obtained from John Stroud, Sept. 30, 2006.

"Wind Energy in Germany 2009." German Wind Energy Association. Copy in author's possession.

WindLogics, Inc. "2006 Minnesota Wind Integration Study." Vol. 2, "Characterizing the Minnesota Wind Resource." Report prepared for Minnesota Public

Selected Bibliography

Utilities Commission, St. Paul, Nov. 30, 2006. http://www.windlogics.com/wp-content/uploads/2012/04/2006-Minnesota-Wind-Integration-Study-vol-II.pdf.

"World Carbon Dioxide Emissions Data by Country: China Speeds Ahead of the Rest." Jan. 31, 2011. Guardian, http://www.guardian.co.uk/news/datablog/2011/jan/31/world-carbon-dioxide-emissions-country-data-co2/.

SECONDARY SOURCES

Asmus, Peter. *Reaping the Wind: How Mechanical Wizards, Visionaries, and Profiteers Helped Shape Our Energy Future.* Washington D.C.: Island Press, 2001.

Cravens, Gwyneth. *Power to Save the World: The Truth about Nuclear Energy.* New York: Alfred A. Knopf, 2007.

Gertner, Jon. "Atomic Balm?" *New York Times Magazine,* July 16, 2006, 62, 64.

Gipe, Paul. *Wind Power: Renewable Energy for Home, Farm, and Business.* White River Junction, Vt.: Chelsea Green, 2004.

Mason, Dr. V. C. "Wind Power in Denmark." *Country Guardian.* Last modified March 2007. http://www.countryguardian.net/.

Righter, Robert W. *Wind Energy in America: A History.* Norman: Univ. of Oklahoma Press, 1990.

Williams, Wendy, and Robert Whitcomb. *Cape Wind.* New York: Public Affairs, 2007.

Selected Bibliography

Index

Index

Illinois, 43, 74, 76, 88, 116
Iran, 15, 18
Iowa, 9, 73, 74, 81, 116, 117
Iowa Utilities Board, 75
Iowa Wind Energy Association, 75
Iraq 15, 147
Israel, 18
Italy, 26, 98

Jacobs, Marcellus, 16
Jacobs Wind Electric Company, 16, 17
Japan, 24, 33, 67, 80, 92, 95–99
    Kyoto, 24, 120
Johnson, Jenny, 6

Kansas, 74, 76
Kaplan, Matt, 80
Kenetech, 22
Kennedy, Senator Teddy, 84, 112
Kentucky, 22
Kerry, Senator John, 119
Kleisner, Ted, 53
Kjaer, Christian, 31
Kuhn, Dr. Thomas H., 124
Kulongoski, Governor Ted, 120
Kyoto Protocol, 25, 30, 40, 41, 96,147

Lake Erie, 81
Lake Ontario, 81
Las Vegas, 101
Las Vegas Sun, 103
Lawrence Berkeley National Labora-
    tory, 134
    People, 52–54, 59, 64, 123
Lewisburg City Council, 61
Lieberman, Senator Joe, 119
Light, Tabatha, 61
Lindzen, Richard, 68
London Array, 39, 76
London Telegraph, 35

Maine, 80, 124
Manchester, John, 53
Manchin, Governor Joe, 60
Many, Floyd, 112
Maple Ridge (wind farm), 81, 110
Marcellus Shale, 91, 138
Marriott Hotels, 22
Martin Next Generation Solar Energy
    Center, 99
Maryland, 6, 117, 123
    Greenbelt, 124
Masdar, 39
Mason County (WV), 88
Mason, V. C., 28, 29
Massachusetts, 81, 85
Massachusetts Institute of Technology,
    68
Maynard, Elliott "Spike," 69
McBee, Henrietta, 1–3, 7, 83
McCullough, David, 84
McKenney, Glenn, 52
McKinney, John, 55
MeadWestvaco, 7, 43, 53–54, 59, 142
Merkel, Chancellor Angela, 33
Merrill Lynch, 20
Methane, 105
Meyer, Glitzenstein, and Crystal, 128
Michaels, Patrick, 68
Mid-Atlantic, 43, 51,61,63, 69,
Mid-Atlantic Highlands, 124
Mideast, 15
Midwest, 2, 9, 16, 68, 72, 73, 77, 81,
    145, 146
    Transmission, 80, 113, 114
Midwest Independent System Operator
    (MISO), 75
Midland Trail Scenic Highway Associ-
    ation, 61
Miliband, Energy Secretary Ed, 37, 38
Minerals Management Service, 85
Minnesota, 17, 72, 75, 116,

Index